The Religious Roots of Longevity Risk Sharing

Moshe A. Milevsky

The Religious Roots of Longevity Risk Sharing

The Genesis of Annuity Funds in the Scottish Enlightenment and the Path to Modern Pension Management

palgrave
macmillan

Moshe A. Milevsky
Toronto, ON, Canada

ISBN 978-3-031-62402-5 ISBN 978-3-031-62403-2 (eBook)
https://doi.org/10.1007/978-3-031-62403-2

Cover credit: Andrea Pistolesi

This Palgrave Macmillan imprint is published by the registered company Springer Nature Switzerland AG
The registered company address is: Gewerbestrasse 11, 6330 Cham, Switzerland

*"Jehoiachin changed from his prison garments, and he ate bread regularly before the king all the days of his life. And as for his provisions, there was a regular ration given him by the king, a portion for each day, **all the days of his life.**"*

2 Kings 25:29-30
New King James Version (NKJV)

*"The fear of the **Lord prolongs life**. The years of the wicked will be shortened."*

Book of Proverbs, Chapter 10, Verse 27

*"Among the several purposes of a society, one should try to arrange for … a **fund** out of which the members may be effectually helped in their needs … not only in the cases of accident, but also in sickness, **old age** and distress."*

Pope Leo XIII
Rerum Novarum, St. Peter's in Rome
15 May 1891

Preface

For generations, the promise of life annuities—fondly referred to as the biblical "daily bread for life"—has been extended to the aged, the widowed, and the retired. These vows of perpetual support, made with the best of intentions, have too often fallen prey to uncertainty, leaving many with promises unkept and the bread undelivered. One of the major challenges was the uncertainty surrounding the longevity of beneficiaries' lives, which made it difficult to budget for random financial obligations. How does one effectively manage longevity risk?

This book delves into the history of attempts made to manage the risk of living a long and unpredictable life. The book's first part explains what longevity risk is, how pension funds and annuity pools can mitigate this risk, and the rather surprising role of religion in that process. The second part—the core of the book—goes back in time to the Church of Scotland's successful implementation of a longevity risk pooling scheme in the eighteenth century. The purpose of this scheme was to provide pensions to widows of ministers and professors. The economist Adam Smith, who taught at the University of Glasgow, was among the notable figures of the Scottish Enlightenment who participated in this scheme.

Many previous authors have praised the Scottish scheme as a revolutionary development in actuarial science, probability, and statistics. They have credited mathematicians and emerging scientists for spearheading the initiative. Some have even gone so far as to claim that the eighteenth-century scheme was a victory of science over superstition. However, a closer look at the archival records—the main scholarly contribution of this work—reveals a more nuanced account. The documentary evidence suggests that Protestant beliefs, individuals, practices, and institutions played a vital role in developing best practices for managing longevity risk. I argue that the eighteenth-century financial engineers and trustees were devoutly religious individuals—including the main driving force behind the scheme, Reverend Alexander Webster—many of whom believed mortality rates followed "divine" probabilities and that longevity modelling was another branch of theology. Moreover, I believe that religious faith is what provided the confidence to make century-long financial projections for the value of fund, which bordered on prophecies.

The book's third and final part explores some of the challenges that big, impersonal longevity risk-sharing pools have faced going into the twenty-first century. Nowadays, traditional kinship and affinity ties are lost, which has raised concerns about pension equity. In fact, there is increasing evidence that mortality and longevity rates differ depending on socio-economic status. This prompts the question: if the wealthy are expected to live longer and receive pensions for an extended period, should the poor be the ones asked to "pass the plate"?

Toronto, Canada
20 June, 2024

Moshe A. Milevsky

Acknowledgements

Throughout the two years I have been working on this project—starting with my very first visit to the archives at the General Register House on 2 Princes Street—I have been concerned with the optics of *cultural appropriation.* As a middle-aged, white orthodox Jewish male from Toronto, Canada—who can't even try haggis—I'm not the ideal person to tell the Presbyterian story of an eighteenth-century innovation to assist widows and orphans, managed by the Church of Scotland. Indeed, that could have been a task for Walter Scott, Robert Luis Stevenson, or Arthur Conan Doyle.[1]

So, to those readers (and reviewers) who might be wondering the exact same thing, I'll admit a certain level of discomfort in immersing myself and writing about this topic. Nevertheless, while I lack the Scottish background, I have approached this undertaking from (1) a distinctly financial and actuarial perspective and (2) with a religious upbringing and seminary training, both of which I *can* claim some cultural expertise.

More importantly, this book is an expanded version of my (2022) MSc thesis at the University of Edinburgh's Department of History, Classics and Archeology, where I studied during a sabbatical from York University. So, I would like to begin by acknowledging my MSc thesis supervisor, Dr. Alasdair Raffe, for his guidance, as I struggled to catch-up on a rich

[1] Or, more recently Ewan McGregor, Gerald Butler, Craig Ferguson, Annie Lennox, and Sean Connery (with a ghost writer.)

history that is likely known by most Scottish school children. In particular, Dr. Raffe's post-graduate course on the eighteenth-century Scottish Enlightenment is what triggered my interest in this period. That is where I (also) learned that 1745 should be known for things other than Jacobites or books by Diana Gabaldon.

In addition to faculty at the University of Edinburgh, I would like to thank Emma Filshie, Simon Johnson, and Ruth Jones at the National Records of Scotland (located at the General Register House) for their assistance and help during my many visits to the archives. In addition, I would like to thank the office of the Moderator of the General Assembly of the Church of Scotland for permission to reproduce an image (Fig. 5.3) from the archives.

Finally, I would like to acknowledge Shlomo Benartzi, Mark DiGiorgio, Chris Lewin, Andrew McDiarmid, David Raymont, Stephen Richards, Trevor Sibbett (sadly, recently deceased), and Koichi Tanaka for helpful comments on earlier drafts and related papers. Finally, thanks to Emma Bromley for transcribing the fund trustee minutes, Madeleine Salvadori for phenomenal research assistance during the final stretch of work, and my editor Ellie Duncan and the staff at Palgrave (the publisher).

Contents

Part I A Blessed Old Age

1	**Do You Believe in Pensions?**	3
	The Good Retirement	3
	So, What Factors Drive Pension Rankings?	6
	Why Should Protestants or Religion Matter?	9
	Enter Max Weber	11
	My Historical Theory: Who Got There First?	12
	The Church of Scotland Annuity Fund	14
	Overview and Outline of the Book	17
	Sources and Additional References	21
2	**Longevity Risk and Religion**	25
	Reading the Obituaries	25
	Visiting the Gravestones	32
	Deaths of Despair	35
	Social Versus Religious Components	37
	Clergy Versus the Rest	39
	Longevity: Now Versus Then	41
	Did People Know Their Risks?	42
	A Religious Model for Longevity	44
	Conclusion for a Long Life	46
	Sources and Additional References	47

3 **The Benefits of Pooling** 49
Introduction and Stuff 49
How Much Money Do I Need to Retire? 50
What Exactly Do These Living Probabilities Mean? 55
Couple Plans for Retirement and Pools 57
Three Is a Crowded Pool, But Is Even Better 60
Measuring Economic Utility Benefits 62
Mixed Pools and Mortality Heterogeneity 65
Sources and Additional References 66

Part II Church & University

4 **An Enlightened Financial Innovation** 71
Summary of Part One 71
Introduction to History 72
Caring for Widows Over the Ages 74
The Westminster Confession of Faith 75
The First Mandated Savings Plan 76
Actuarial Historiography 79
Methodist versus Presbyterians 81
Sources and Additional References 82

5 **A Presbyterian Scheme for Ministers** 85
Introduction & Plan 85
High-level Review of the Scheme's Features 87
A Note on Money 90
The Reversionary Annuity: Choices 91
Choice Architecture versus Choices Made 94
A Glimpse into the Archives 95
Where Are These Financial Contributions Material? 97
Revisions to the Founding Text 98
A Protestant Management Ethic 100
Sources and Additional References 102

6 **Alexander Webster and the Archives** 107
Alexander Webster: Administrator Extraordinaire 107
Managing a Meeting of the Fund Trustees 109
We Seek to Avoid "Greatly Embarrassing the Trustees" 113
"Fatuous Person and Idiot without Tutors" 123
Revisions to the Scheme in 1748 124
Choices Made and Contributions Set 126
Conclusion 138
Sources and Additional References 140

7 **Annuity Management in the Eighteenth Century** 145
Was the Mathematics Known and Used? 145
Theoretical Value: Payout Multiples 146
Numerical Examples: Payout Multiples 150
The Annual Financial Statements 152
Sample "Charge and Discharge" 154
Narrative Versus Numerics 155
Garbage-in-Garbage-out 156
"Must Rise to This Amount" 157
From Charges & Discharges to Fund Values 158
"Money Lying Dead" 158
Minister's Arbitrage Versus Usury 163
Conclusion and Takeaways 166
Sources and Additional References 167

8 **From Church PAYGO to Fully Funded** 171
Introduction and Objectives 171
Fasti Ecclesiae Scoticanae 173
Patrick Couper, Pittenweem, and Witches 175
The Earlier Scheme of 1717 177
From Synod Versus Presbytery 179
Daniel Defoe's Path to a Financial Innovation 180
From Small to Large 182
Church Turmoil 182
To Sum Up the Controversy and Timeline 186
Sources and Additional References 188

9	**Scientific Models Versus Religious Beliefs**	191
	Introduction and Objectives	191
	Stochastics in the Eighteenth Century	195
	From Pearson to Stigler to Hald	196
	Variance Ignorance and Minister Price	198
	Conclusion	200
	Sources and Additional References	201
10	**The First Biblical Annuity**	205
	Jehoiachin's Pension Annuity	207
	The Last Living King of Judah	209
	Judean Genealogy	211
	Archeological Evidence and Cuneiform Tablets	213
	Valuation of the Babylonian Annuity	214
	Annuitant Mortality: Borrow from a Roman	215
	Interest Rates in the Fifth Century BC	217
	The Richest Man in Babylon	219
	Conclusion	220
	Sources and Additional References	222

Part III From Retirement to Decumulation

11	**Pension Resistance in the Nineteenth Century**	227
	Introduction & Objectives	227
	Forces: For & Against	231
	Pension Scheme Classification	232
	Pension Origins & Objections	234
	National Examples: Germany, the UK, and the US	236
	Conclusion	240
	Sources and Additional References	241
12	**Longevity Heterogeneity in the Twenty-First Century**	243
	National Pension Pools: Fair IRRs	244
	Wealth Transfers: From Less to More Fortunate	247
	Life Annuities: Quantify the Wealth Transfer	249
	Can the Volatility of Longevity Justify the Subsidy?	252

Should the Short Lived Get Bigger Pensions? 253
The Longevity Gap 256
Ethics of Pension Equity 257
Final Fund Epilogue 259
Sources and Additional References 259

Index 263

About the Author

Moshe A. Milevsky is a professor at the Schulich School of Business, York University (Toronto) where he teaches courses on Wealth Management, Personal Finance, Retirement Modeling, and Pension Planning. He is also a member of the Graduate Faculty in Mathematics and Statistics, where he co-supervises doctoral students, and a (2002) Fellow of the Fields Institute for Research in Mathematical Sciences. He has graduate degrees and training in Financial Economics, Mathematical Statistics, Probability Modeling and (most recently) History from the University of Edinburgh.

Professor Milevsky has published 17 books and over 60 peer-reviewed articles, including in the *Financial History Review*, as well *Journal of Financial and Quantitative Analysis* (JFQA), *Journal of Banking and Finance* (JBF), *Review of Finance* (RF), *Mathematical Finance* (MF), and *Financial Analysts Journal* (FAJ). In addition, he has published many articles on annuities and longevity risk in the *Journal of Risk and Insurance* (JRI), the *North American Actuarial Journal* (NAAJ), and *Insurance: Mathematics and Economics* (IME), where he is a member of the editorial board.

His research interests and expertise are in the interplay between *actuarial insurance* (mathematical modelling) and *financial economics* (decision making), with a particular focus on the decumulation phase of the human lifecycle. His prior book, *King William's Tontine: Why the Retirement Annuity of the Future Should Resemble Its Past* (Cambridge University Press), was awarded the 2017 Kulp-Wright Book Award from the American Risk and Insurance Association.

List of Figures

Fig. 1.1	Pension ranking and religion	9
Fig. 2.1	Visualizing longevity risk and dispersion	29
Fig. 5.1	The Act of Parliament Taking Effect 25 March 1744 (*Image Source* Great Britain (1743 [1744]) Eighteenth Century Collections Online [ECCO])	86
Fig. 5.2	Scheme choice notification forms	91
Fig. 5.3	Glasgow professors and 1744 scheme lists	96
Fig. 9.1	Calculated projection vs. real value	192
Fig. 10.1	Release of King Jehoiachin and his annuity	210
Fig. 10.2	Valuation picture for guaranteed longevity risk pooling	215

List of Tables

Table 3.1	What do you need at age 65 to withdraw $100,000 per year?	54
Table 3.2	The financial benefits of pooling resources	57
Table 5.1	Contribution premiums versus benefit payouts	92
Table 5.2	Scheme choice assumptions versus realizations	94
Table 6.1	Work of trustees: Appearances, attendance, and activity	112
Table 7.1	Theoretical payout multiples at 4% interest	151
Table 7.2	Fund charge and discharge: 1745 to 1755	154
Table 7.3	Scheme fund projections vs. actuals	160
Table 8.1	Timeline of key events: 1718 to 1748	187
Table 10.1	Lifespan and the final kings of Judah	212
Table 10.2	Ulpian's table circa 300 CE	216
Table 11.1	Global pensions heroes and beginnings	230
Table 12.1	Longevity risk pooling: hypothetical winners and losers	250
Table 12.2	Work and contribute $1 per year to pension for 40 years	254

PART I

A Blessed Old Age

CHAPTER 1

Do You Believe in Pensions?

The Good Retirement

We live in a world awash in rankings. Products, services, schools, and even people are routinely and quantitatively tabulated with medals and honours granted to those at the top. And alas, pensions have recently joined this game of scores. So, although national retirement plans have been in existence for over 130 years, initially encouraged by Pope Leo XIII in the earlier quoted encyclical, they now earn letter grades in addition to salvation.

A few years ago, the US-based CFA Institute and the Australian-based firm Mercer joined forces to rank pension global systems. Every year the rankings are updated, and new countries are added to the list. Presumably, a committee of wise women and men meets once a year in a conclave to evaluate the players. The result of that labour and deliberation is one comprehensive summary score of every country's pension system, ranging from a theoretical 0 to a nearly impossible 100. That score is also mapped onto a letter, ranging from A to E. There are no Fs, perhaps because nobody fails when there are grades for effort, similar to how a university or college professor might assign final grades these days.

A national pension scheme—such as the Canadian Pension Plan or Old Age Security (in Canada) or the Social Security system (in the US), for those readers who might be a tad young to care or be aware of these plans, consists of a guarantee or promise by a government or state to pay an

M. A. Milevsky, *The Religious Roots of Longevity Risk Sharing*,
https://doi.org/10.1007/978-3-031-62403-2_1

income to citizens when they retire and "get old." Without getting into a detailed discussion of "pension pillars" or "PAYGO versus funded," the promise of income or support in old age is quite different from welfare benefits or social assistance for the poor.

The retirement pension benefit that citizens are entitled to usually depends on the years they have worked in the country, how much they have earned, and especially how old they are when they decide to retire. The age factor, in particular, is a rather sensitive political topic in many countries, as evidenced by (massive) protests in France during the spring of 2023 when the government attempted to force citizens to retire at the age of 64 versus the age of 62. In fact, oddly enough, the French youth—who were decades away from their 60s—were some of the loudest and most vocal protestors against these reforms, perhaps because they feared having to support their own parents and grandparents. Nevertheless, at the time of writing, the law has passed, and the French must work two more years to earn their full pension, *sacré bleu*!

Back to the pension rankings, it seems that despite their cushy and early retirement, the French don't actually have the best pension system in the world. Nor is it the Italians or the Greeks, who live in a country believed to be a paradise for retirees. Instead, the best national pension system in 2022 was in Iceland. Yes, that is a small island country of glaciers, geysers, and lava fields with less than 400,000 inhabitants who are mostly descendants of Vikings. They have the strongest pensions on the planet. No, it certainly isn't the country I would have guessed either and it sounds rather "random," as a millennial might assert. Moreover, this wasn't a year 2022 fluke. In the next year's ranking, that is for the 2023 competition and the most recent at the time of writing, Iceland ranked in the number two spot as well. Why? I'll get to the reasons later.

Moving down the ranking for 2022, the number two (a.k.a. silver medal) spot was a country not far away geographically, Denmark, and rounding out the top three was the Netherlands.[1] Other countries ranked in the top ten were Finland and Norway, as were Australia and Israel. So, these awards aren't backhanded prizes for the Nordics and their largely socialist economies. In fact, 18 countries worldwide earned good grades, ranging from A to B, and they are scattered all over the globe. Now,

[1] In the year 2023, the ranking was reversed and the Netherlands was number one, Iceland was number two, and Denmark was number three. Not much change from year to year, since pension systems are large and complex undertakings.

moving to the so-called losers in the year 2022, at the bottom of the list were the countries of Türkiye (a.k.a. Turkey), Thailand, Brazil, Argentina, Poland, and Italy. They all earned Ds and Es. Remember, this is a ranking of national pension systems, not soccer prowess or FIFA rankings, where obviously Brazil and Italy and Argentina would be at the very top. No different from other rankings, which are updated yearly as noted earlier, the grades do tend to shift from year to year, but similar to national soccer teams, there aren't any sudden changes.

One thing is for certain, this particular competition might not get the publicity of the FIFA World Cup, but the winners do earn and benefit from Andy Warhol's 15 minutes of fame and publicity. For example, in November 2020, when the Netherlands and Denmark won top scores, the news and national press releases immediately appeared in the venerable *Financial Times*.[2] The next year, when Iceland's national pension system was crowned the best and India's the absolute worst, it made headlines in both countries.[3] In fact, pension industry executives and government regulators around the world routinely boast and tout their Mercer CFA Pension Index grade or lament their methodological shortcomings if their country happened to tumble in recent tables. In other words, these grades matter to many people, other than pension nerds and geeks. The point is that the Mercer CFA Institute Global Pension Index—to use the full name—has industry gravitas, even though other global organizations have developed alternatives.[4]

Just to be clear once again, for those who might suspect I'm cherry-picking years, per the most recent data available from 2023, the top three spots were secured by the Netherlands, Iceland, and Denmark. These, as noted, are the same countries that won in 2022 but in a slightly different order. Thailand, Türkiye, India, Argentina, and the Philippines were awarded the lowest grade of D in the pension category. Compared

[2] Pandemic will hit the pension prospects of billions, warns study. *Financial Times*, 19 October 2020 (see https://www.ft.com/content/ed619409-dab9-49fb-bef4-2b5824 07353d).

[3] *The Business Standard*, 20 October 2021 (see https://www.business-standard.com/article/international/iceland-has-best-pension-system-in-the-world-india-near-bottom-of-index-121102000032_1.html), and *The Times of India*, 20 October 2021 (see https://timesofindia.indiatimes.com/business/international-business/worlds-best-and-worst-pension-systems-in-2021-all-you-need-to-know/articleshow/87160466.cms).

[4] See, e.g., the Paris-based OECD ranking and pension statistics (https://www.oecd.org/finance/private-pensions/globalpensionstatistics.htm).

to 2022, Poland and Italy managed to move up from a D to a C, though the difference in points was minor. Also, like FIFA that adds more teams to the tournaments as time goes on, in 2023, Botswana was introduced into the pension rankings, with a grade of C, along with Croatia and Kazakhstan, both of which earned C+. Brazil, Mexico, and Peru, on the other hand, only managed to earn Cs. So, there were no big changes or surprises, compared with 2022.

Now, before I delve into what drives good pension grades—which gets to the heart of this chapter and the core of this book—I should emphasize that these grades and rankings and scores aren't solely about the amount of income promised to retirees, or money in the bank, or how well their investment asset allocation performed in a particular year. That would be a very narrow and one-dimensional proxy for the health of a country's pension system. Rather, the Mercer CFA Institute Global Pension Index score takes into account myriad environmental, economic, as well as retirement-related factors. If you look at the fine print or details of the methodology, 40% of the weighting score is allocated to measures of system adequacy, that is, how generous the retirement benefit is for old people; 35% weighting is given to broad measures of sustainability; and 25% to measures of integrity—ensuring countries will be able to fulfil their generous financial promises. By promises, I of course mean the guarantee to pay a lifetime of income to a beneficiary, a.k.a. a life annuity or simply annuity. The ranking is broad, comprehensive, and all encompassing.

So, What Factors Drive Pension Rankings?

Although the ranking of national pension systems is clearly a complex and multifaceted process, after reviewing the methodology I couldn't help but wonder if there was just one or two specific factors that had the most meaningful impact on the final grade. I asked: *Is there some secret to a great (country) pension*? Yes, there is always a risk of simplification and summaries, but perhaps this is no different from Lionel Messi being the single most important factor in Argentina's soccer success (OK, perhaps with Julian Alvarez as a strong second), even though the entire team hoisted the World Cup.

So, I decided to "kick around" the data and numbers to examine if there were any noteworthy or unexpected factors driving the pension winners and losers. As someone who also works as a consultant in the retirement and pension industry, I was searching for some factor or lever that was critical to a good grade. For example, when I saw the Nordic countries ranking highly, I first thought it was the northern latitude, their cold and snow, but that wouldn't explain Canada's B. And, that couldn't really be "optimized" or tinkered with either. In fact, even if there was a link between cold white snow and pension health, and even if one could do something about the weather, it wouldn't indicate causality—which is something that professors drill into every student in Statistics 101.

Moving on in search of some hidden factor, I thought perhaps the life expectancy of the population, which is a statistical measure of how many years a typical person is expected to live, might be an important factor. I'll delve into life expectancy and how long people live in the next chapter. I was thinking that perhaps the longer people live, the more retirement income they require, which necessitates a better national pension system, and so on. It seems that yes, that factor was an important player on the team. Technically, when I regressed national pension index scores on national life expectancy, I did find a positive link (a.k.a. correlation), but a weak one. I then performed the same exercise, akin to measuring correlations with country fertility rates, and the total number of births per woman. There, too, I found some significance, which is not surprising if you consider that more babies and future workers can help sustain a retirement pension system.

And so—getting to the core of my story and message—as I continued to search for one (perhaps mythical) hidden factor that could perhaps predict or explain how high or low a country's national pension system would rank—call it the Messi factor—I stumbled across something rather unexpected and even "random." It appears that a country's pension score is highly correlated with—get ready for this—their *religion*.

No, this isn't a story about vague religious beliefs, mysticism, and superstition. The best factor for predicting the pension index score—from Iceland's top 84% down to Thailand's 41% and near-failing grade—was the fraction of a country's population that identified as *Protestant Christian (PC)*, versus Catholic or Muslim, or Jewish. The higher the PC percentage, all else being equal, the higher the index score. How did I get that PC number? The Protestant Christian (or just Protestant) fraction was obtained from the Pew Research Center and confirmed against data from the Cline Center for Advanced Social Research. The PC fraction ranged from Iceland's 91.3% (highest), that is, 91.3% identified as Protestant Christians, while the other 8.7% of the population of Iceland identified as being part of another religious group, or no group at all.

At the very other end of this PC spectrum is Thailand, with 0.4% of the population identifying as Protestant Christian, which was the lowest PC number in the entire ranking. For the record, Buddhism is the largest national religion in Thailand.

Recall that the Protestant Reformation was a process that began in Germany in 1517, when Martin Luther published (and purportedly nailed to the door of a church) his Ninety-five Theses criticizing the Catholic Church. Fast forward 500 years, by which time his "new religion" has spread around the world, with some "troubles" along the way. By the early twenty-first century, there were a billion Protestants—yes, in many denominations, and I'll return to that—scattered around the planet. To my point, the fraction of a country's population that identifies as Protestant Christian is a strong predictor of that country's pension index score. The higher the PC, the better the pension, as shown in Fig. 1.1, displayed nearby,[5] in which the horizontal x-axis represents PC intensity, and the vertical y-axis is the pension index score. The positive sloping line is the smoking gun in regression as long as the estimated coefficient is statistically significant. Even after controlling for the life expectancy and fertility rates, the *t*-statistics value was close to 6, above the 5% significance.

This result is suggestive and interesting but not the only reason that a country's pension system is ranked higher.[6] Warning: statistically speaking, this linear relationship that you see in Fig. 1.1 is far from perfect, and there are many exceptions to this rule. For example, Israel ranks very highly (see the dot near the number 75), but it doesn't have that many Protestant Christians. In Israel, approximately 73.8% of the population is Jewish, 18% Muslim, 1.9% Christian, and 1.6% Druze. Now, I am obviously not suggesting that a country change its national religion to improve its pension rankings. Just as importantly, it could very well be that the Protestant Christian fraction is a proxy for another hidden variable. But you must admit, this is rather odd and interesting, no?

So, notwithstanding my disclaimers and the possibility of confounding factors, the next question you might ponder is why? What is the link

[5] Created and generated by the author.

[6] A full and proper discussion of this result is provided in a co-authored research paper with Marcos Velazquez, entitled "Protestants and Pensions" (www.ssrn.com). In fact, we find that American states and regions that are more Protestant are more likely to have better-funded pensions compared to those that lean Catholic. This isn't just a national story. See also Friedman (2021) as well as the work by Grünewald (2021).

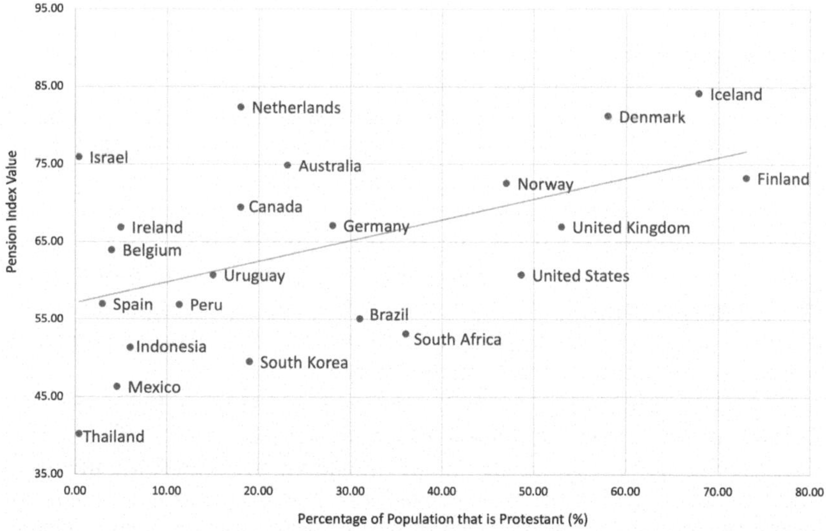

Fig. 1.1 Pension ranking and religion

between Protestantism, its religious beliefs, or its practices, on a pension system's adequacy, sustainability, and integrity in a very secular twenty-first century? Is it just a snowy fluke—remembering that correlation isn't causation? Or is there something deeper going on here? That finding is what sent me on an intellectual journey of discovery and led to much of the research reported on in this book.

Why Should Protestants or Religion Matter?

Not all readers might be surprised to see a link between the Protestant Christian fraction and pension rankings, or between religion and finance. In fact, there is a (very) large body of research literature that examines the impact of religious beliefs on financial and economic outcomes. The literature is vast and some articles in the field of "Economics of Religion" have been cited thousands of times. For example, and to our point, one article in the journal called *Oxford Economic Papers* documented, using survey data, that Protestant Christians generally exhibited a greater sense of financial responsibility in their economic attitudes.[7] And so, better pensions might be a manifestation of that sense of responsibility. Another article that was published in the *Review of Economics and Statistics* found—this one using laboratory experiments—that Protestant (versus Catholic) Christians were more likely to increase contributions to public goods after being primed to consider their religious views. Those are just two of many studies that link Protestant Christian beliefs and affiliation to tangible economic outcomes.[8]

In other words, there are noticeable financial and economic differences between Catholic and Protestant Christians that might flow-thru and influence retirement financing and policies.[9] So, it isn't completely unexpected (or random) to see such differences.

Not to belabour the point, but in yet another widely cited paper that appeared in the *Journal of Financial Economics*, the authors found that geographic regions with more Catholic relative to Protestant investors tended to invest in lottery-type stocks, which is consistent with a more favourable religious attitude towards gambling among Catholic Christians.[10] Perhaps Protestant Christians don't like or want to "gamble" on their financial future and therefore have acted to set up—or lobbied politicians to implement—better national pension systems? Yes, this is speculative but the explanation is consistent with the pension rankings.

With all that in mind, perhaps one potential explanation for why Protestant Christian countries rank higher is because of the relationship between Protestant and Catholic attitudes and psyche, versus theology and beliefs per se. Indeed, researchers have established that religious affiliation plays a critical role in determining an individual's perspective on solidarity, traditionalism, and—quite importantly—the responsibility of children to support their ageing parents. The idea is that individuals who either adhere to Catholic Christian beliefs or individuals who are part of a Catholic Christian community are more inclined to financially assist

[7] Renneboog and Spaenjers (2012).

[8] In the earlier noted working paper (with M. Velazquez) "Protestants and Pensions" (www.ssrn.com), we provide a more comprehensive review of the literature.

[9] Benjamin et al. (2016).

[10] Kumar et al. (2011).

or support their parents—or at least believe that they should do this—during their retirement and older years relative to Protestant Christian beliefs and/or communities.

One of the very early studies in this field—conducted over 80 years ago, well before national pension systems were well established or ranked—suggested a "[reasonable] association between religious affiliation on the one hand, and opinion regarding the obligation of children to give support [to parents]."[11] In that study, Catholic Christians tended to adhere most strongly to the belief that children should give support to older parents, while Protestant Christians exhibited the most contestation—or disagreement with that statement—relative to other religious affiliations. In other words, an above-average number of Protestant Christian affiliated children did not feel an obligation to support their parents, and their parents did not expect to be supported. This is unlike Catholic Christian (guilt-laden?) children who did feel an obligation to do so.

In fact, when I was describing the results of my nascent research at a conference in Rome (Italy) a few years ago—where you'll recall the pension ranking is quite low and most of the population is Catholic Christian—that was exactly the (informal, non-scientific) reaction I received from the audience. Italians felt an innate obligation to assist, take care of, and help their elderly parents. In fact, some were shocked to hear that any child might feel otherwise. I argued that Protestant adults worry that their children might abandon and discard them (to be very blunt) in old age, they need to rely on themselves, which encourages them to set up "good" national pension systems.

Enter Max Weber

Another more sociological and economic-based explanation, although not entirely distinct from a psychological one, can be traced to the work of the late nineteenth-century sociologist Max Weber. In Weber's book *The Protestant Ethic and the Spirit of Capitalism* (1902), he observed that in his home country of Germany, "business leaders and owners of capital...tend to be predominantly Protestant [relative to Catholics]," and that "the reasons for these differences must be sought principally in their

[11] The full study is cited as Dinkel (1944) in the bibliography.

distinct internal characteristics and not in the external historical and political situation of different denominations." Weber's thesis was echoed by R.H. Tawney in his book *Religion and the Rise of Capitalism* (1926). The Weber-Tawney thesis is that Protestant Christian values and practices foster a more positive approach to capitalism, which might (somehow) transmit itself to better national pension systems.

I'll admit that this is vague and controversial since "anti-Webers" have complained that his thesis isn't testable and/or it was refuted by the existence of Italian Venetian (i.e., very Catholic) capitalism in the eighth century, centuries before Luther. Nevertheless, the economic implications of religious differences—either beliefs or practices—between Catholic and Protestant Christians have recently been rekindled by Benjamin M. Friedman in his (identically titled) *Religion and the Rise of Capitalism* (2021). My point here is that there is a large body of scholarly literature written by sociologists and economists that study the economic and financial differences between Protestant and Catholic Christians, and national pension systems might be a corollary or by-product of something larger than attitudes towards old age and retirement.

My Historical Theory: Who Got There First?

Not to downplay the importance of psychology, sociology, or economics, I have another theory—an idea that is the launching pad for the remainder of this book—explaining why Protestant Christian countries might have created better pension systems. Back to the soccer analogy, think of countries like Uruguay (where my father grew up) versus Canada (where I live today). The former country has had a competitive soccer team for almost a century, and it won the first World Cup in 1930. Uruguayans have a long history of playing (good) soccer; it's in their blood. Canadians, in contrast, first played in the World Cup in 1986, never won a World Cup, and, despite some recent advances, don't have a history of soccer. They haven't been at it for very long. History, or at least the past, matters for soccer rankings. Yes, I know this is a stretch—at this early juncture—but perhaps the same applies to pension rankings.

In fact, as most pension observers are well aware, the German chancellor Otto von Bismarck is credited with inaugurating the world's first national pension scheme in 1889. He was a (very) Protestant Christian and spent the early part of his professional life struggling (unsuccessfully) against German Catholics in the *Kulturkampf* of the 1870s. Eventually, his design for social insurance was copied by other Protestant—and only later Catholic—countries, reaching the United States in the 1930s. This isn't the place to delve into the history of the German pension system, but national comprehensive age-based pensions started in a Protestant Christian country.

In fact, the first private sector (non-government) pension system for American teachers and professors was created in 1905 by the industrialist Andrew Carnegie.[12] The Teachers Insurance and Annuity Association–College Retirement Equities Fund (TIAA–CREF) retirement scheme offered the first life annuity fund for teachers in the United States. Andrew Carnegie was one of the four founding American tycoons, alongside John Rockefeller, Jay Gould, and J.P. Morgan; all Protestant Christians as a matter of fact.

I now arrive at the launching pad for the core idea of this book. I believe the origins of successful longevity risk pooling—which is the most critical ingredient for pensions—go farther back. Yes, national pensions began in the late nineteenth century in Germany—and I'll return to those birthing pains in Chapter 11—and then eventually spread all over the world. But the idea of "pooling" a large group of workers together, forcing them to make small ongoing contributions during their working life, and, in exchange, offering them deferred life annuities—either to the workers themselves at some advanced age or to their surviving beneficiaries—and managing this successfully, began over a century earlier. The origins or the birth of successful longevity risk pooling, I would like to argue, took place in a church in Edinburgh, Scotland, in the mid-eighteenth century. If one wants to select an actual birthday, it was nearly 280 years ago to the date of writing, in March 1774, give or take a month.[13]

[12] Dale Carnegie was a Scottish Presbyterian, a fact that is more than a trivial curiosity in light of what follows.

[13] Great Britain (1744). It received Royal Assent as 17 Geo. II, c.11 on 2 March 1744, and then went into effect 25 March 1744. This bill was promoted in the House of Common and Lords in late 1743.

The Church of Scotland Annuity Fund

The establishment of the Church of Scotland traces back to the Scottish Reformation in the sixteenth century, marking a significant shift in the country's religious landscape. Initiated by key figures like John Knox, who was heavily influenced by the Protestant Reformation in Europe, the movement sought to reform the church's practices and align them more closely with Protestant principles. In 1560, the Scottish Parliament embraced these reforms, effectively establishing the Presbyterian system of church governance and rejecting the authority of the pope. This pivotal moment laid the groundwork for the Church of Scotland, distinguishing it from the Roman Catholic Church and aligning it with Reformed theology. The church's foundation was rooted in the desire for a church structure that emphasized equality among its ministers, governed by presbyteries rather than a hierarchical episcopate.[14]

Over 180 years later, in March of the year 1744, the Church of Scotland launched an innovative insurance and investment fund, for ministers of the church, although it eventually allowed—for reasons to be explained—professors at the major Universities in Scotland to participate in this scheme as well. This was a period known to historians as the Scottish Enlightenment[15] and many of the literati of the period—and I'll get back to many of them later—participated in this scheme.

Under this scheme, Presbyterian ministers and university professors (i.e., the working participants) made ongoing financial contributions of a voluntary magnitude during their lifetimes, which upon their demise entitled their widows to a life annuity or their orphans to a lump-sum death benefit. So, for example, a minister or professor might contribute 5 "units" per year until he died, and in exchange, the widow would receive 20 "units" per year until she died. Technically, this is called a reversionary life annuity, or simply reversionary annuity. I'll get to the exact parameters and so-called units later on in the book, specifically in Chapters 5 and 6. But, this really was also the first mandated-by-law employer pension scheme, with "default" options and "auto-enrollment" features.

[14] See Ryrie (2006), for much more on this period, which can hardly be summarized in one paragraph.

[15] For more on the period known as the Scottish Enlightenment, and the intellectual giants of that era, see Sher (2015), as well as the book by Buchan (2003).

The 1744 Presbyterian scheme for widows and orphans—I'll discuss the orphan's benefits later—was managed by the Reverend Alexander Webster (b. 1707, d. 1784), designed by the Reverend Robert Wallace (b. 1697, d. 1771), and with technical assistance provided by University of Edinburgh Professor of Mathematics Colin Maclaurin (b. 1698, d. 1746), a student and disciple of Isaac Newton. While I'll get back to each one of these three Scotsmen later on in this book—and, in particular, I'll emphasize the role of Alexander Webster whom I consider to be the protagonist of this book's underlying story—I can state here the very obvious fact that they and the Presbyterian Church were all (very much) Protestant Christians.

Why was this scheme[16] so important? What was unique about this historical scheme? Well, to start with, this reversionary annuity investment fund was the first mandated scheme of its kind,[17] established by an Act of Parliament, although limited synod-based schemes had been operating as early as 1718, by the Church of Scotland, as I will discuss in Chapter 8. Those early attempts did not involve much "longevity risk pooling" and certainly couldn't be considered successful.

Now, for the introductory purposes of this opening chapter, although you personally might not have ever heard of the 1744 Presbyterian Church's reversionary annuity fund, this isn't a revelation or some archival discovery. I should note that much praise has been heaped on this Scottish scheme, especially by actuaries of Scottish origin. I'll do a proper literature review in Chapter 4, and discuss the varied opinions about the scheme's historical importance. In particular, many actuaries view the year 1744 with the same reverence as the year 1693, which is the year when the astronomer Edmond Halley (b. 1656, d. 1742), a contemporary of Isaac

[16] The word scheme (in the twenty-first century) has negative connotations, but that was the word used in the documentation of the eighteenth century, and for the most part, I will stick to that terminology, although occasionally refer to it as the reversionary annuity investment fund (yes, a mouthful), or just fund.

[17] Without prejudice to the Royal Navy's Chatham Chest in 1590 and the larger scheme created by James, Duke of York in 1672, neither of which mandated participation by an Act of Parliament nor was as broadly based. Another related scheme for clergy of the Church of England, was Queen Anne's Bounty. See the older article by William Richard LeFanu, available from the Online Internet Archive (https://archive.org/stream/queenannesbounty00lefauoft/queenannesbounty00lefauoft_djvu.txt).

Newton, published the first empirically based valuation of life annuities.[18] To insurance actuaries, the 1744 scheme and associated technical documentation is considered an important scientific milestone located between Abraham de Moivre's (b. 1667, d. 1754) book *Annuities Upon Lives* (1725) and Richard Price's (b. 1723, d. 1791) book *Reversionary Annuities* (1771). I'll return briefly to Halley, de Moivre, and Price in later chapters, since they are all part of the story of "longevity risk pooling." The bulk of the action was based in the city of Edinburgh, which at the time was the financial centre of Scotland.[19]

In fact, admiration for the Scottish scheme has travelled beyond Scottish mathematicians and actuaries, or members of the Presbyterian clergy who are quite proud of this history. The famed historians and public intellectuals Yuval Noah Harari and Niall Ferguson, writing separately, denoted the year 1744 as a human milestone in the transition from religious superstition to scientific modernity.[20] Harari (2014) in *Sapiens* writes that in seeking guidance on how to design the scheme, the Reverends Alexander Webster and Robert Wallace "did not pray to God to reveal the answer. Nor did they search for an answer in the Holy Scriptures or among the works of ancient theologians," but instead consulted with Mathematics Professor Maclaurin to establish the fund on a sound statistical basis. Again, the existence and importance of this scheme are rather well-known.

Ferguson (2008) writes in *The Ascent of Money: A Financial History of the World*, that "modern actuaries still marvel at the precision with which [Webster and Wallace] did their calculations" (p. 193). I'll get to that "marvel" in Chapter 7, but clearly both Yuval Harari and Niall Ferguson—who were writing very general world histories—were impressed enough with the scheme to give them a position of prominence in their book.

[18] For more on Edmond Halley and his contributions to retirement and annuity planning, see Milevsky (2012), *The Seven Most Important Equations for Retirement*.

[19] The relevant scholarly literature is Bruneau (2020), Deuchar (1896), Dow (1975), Dunlop (1992), Hald (2003), Hare and Scott (1992), Lewin (2003), and Mackie (1956). See also Bellhouse (2011) for a lovely biography of Abraham de Moivre. For more on the City of Edinbirgh, and a very brief discussion of the scheme, see Perman (2019).

[20] Both Harari and Ferguson confuse the 1744 Church of Scotland widow's scheme for ministers and professors with the completely unrelated insurance company Scottish Widows, founded in 1815 to care for widows of soldiers in the Napoleonic Wars.

Now, whether or not the 1744 scheme represented a scientific "big bang" or a manifestation of deep religious beliefs and Protestant practices is a key part of the story and will be properly addressed later in this book. So, while I agree with Harari and Ferguson about the importance of this annuity scheme, managed by Presbyterians, which is a precursor to modern pensions and perhaps a reason why Protestant Christians are likely to have better pensions, I take issue with one important aspect of their arguments.

In a nutshell, these cheerleading chroniclers made two errors. First, they underemphasized the Presbyterian Church's governance practices, administrative infrastructure, moral culture, and Presbyterian beliefs. Second, they neglected to credit the earlier synodical (smaller, regional) origins and the evolutionary process by which the 1744 scheme came into existence. I'll return to all these arguments in Chapters 8 and 9. But the key takeaway at this point—which I suspect Harari and Ferguson would agree on—is that perhaps Protestant Christian organizations had more experience, institutional infrastructure, and practice in setting up pension annuity schemes. But I will go one step further than Harari or Ferguson and argue that Protestant Christian beliefs were just as important as well. Perhaps those beliefs and practices enabled Alexander Webster and his colleagues to harness the nascent field of probability theory to manage longevity risk. My argument in a nutshell is that these Protestant-based factors created a culture, an environment, and a fiscal temperament for pension planning in the mid-eighteenth century.[21] This is the religious origin of longevity risk sharing and pooling.

Overview and Outline of the Book

To be clear, and going back to the motivation and early part of this chapter, the objective of this book is *not* to convince you beyond a reasonable doubt that Protestant Christian countries have better pension systems, or even to completely explain why this might be the case. It is ridiculous, silly and simplistic to pin national pension plan success and failure on one factor, let alone religion, in the twenty-first century.

[21] To be clear, not all contributory insurance schemes managed by Protestant churches in the eighteenth century were successful, and the Methodist's Preacher's Fund, discussed by Norris (2017) is a good example.

Rather, the higher pension ranking of Protestant countries is the *motivation and departure point* for (1) a discussion of the 1744 annuity scheme, and (2) the historical argument at the core of this book. Namely, Protestant Christian beliefs and practices played a critical role in the mathematical, statistical, and actuarial foundations of successful longevity risk pooling, which is absolutely critical for modern pension management. The lynchpin of my argument is that the very first funded pension annuity scheme in the mid-eighteenth century was designed by Protestant ministers and thought leaders, employing Protestant practices and managed by the Presbyterian Church in Scotland. This book will argue that Protestant Christians "got there" first and were able to achieve successful lift-off—as a result of their religious institutions—and therefore it's quite possible that remnants of those origins, like Neanderthal DNA, are still embedded in today's retirement plans.

There is much material to cover, and first and foremost I must begin by providing a clear and yet not too technical explanation of longevity risk and pooling. So, in the upcoming Chapter 2, I will explore the concept of longevity risk, which refers to the uncertainty of people's lifespan during their retirement years. Additionally, I will delve into the role of religious affiliation in the prediction and forecast of longevity. You may be surprised to discover the connection between religion and longevity. After understanding longevity risk, Chapter 3 will explain the benefits of "pooling" longevity risk, which involves grouping and sharing risk with others. This approach has the magical effect of increasing investment returns and reducing overall economic risk. Think of it as diversifying your life but with others. The chapter will also touch on questions of equity and fairness, which will be addressed in greater detail (as it relates to the twenty-first century) in the final chapter (Chapter 12) of this book. Overall, the first three chapters constitute **Part I** of this book, which focuses on more of the technical aspects of longevity risk, and not as much on history or the 1744 scheme. In some sense, the first three chapters are a very elementary Actuarial Science 101 course.

With that context and background in place, **Part II** of the book is where the historical part begins, going back in time and delving into narrative history (*what exactly happened?*) as well as historiography (*how have people written about it?*), of the first successful attempt to tame and manage longevity risk in the eighteenth century.

Chapter 4 of the book, which is the first chapter in Part II, sets the tone for the entire second (historical) part of the book. In particular, it discusses how prior authors, scholars, and historians have written about the eighteenth-century Presbyterian scheme for widows and orphans. In that chapter, I make the case that (some) prior authors may have missed key aspects of the 1744 scheme's religious roots and origins, while others and experts on the Scottish Enlightenment might have ignored the scheme all together and in particular the key role of (our protagonist) the Reverend Alexander Webster. Chapter 5 finally gets into the nitty–gritty details of the mechanics and operations of the scheme, including the exact amount that was charged to participants, when it was paid, how much was due to widows and orphans, and so on—the comparison and contrast with modern (twenty-first century) retirement plans is inevitable. Chapter 5 also explores the "choice architecture" as well as the investment decisions that participants were asked to make, the choices they actually made, and whether the burdens—savings rates—they imposed on participants were substantial. What were typical wages or salaries for clergy and professors?

Moving on, Chapter 6 is the archival core and main historical contribution of the book. The content in that chapter is based on the handwritten minutes of the trustee meetings during the first few years of the scheme, from mid-1744 onward. It details the often week-by-week struggles and challenges the trustees faced in managing the scheme and in particular speaks to the importance of individual ministers within the Church of Scotland in ensuring the scheme's survival. This chapter provides the evidence that pension and annuity fund success goes well beyond implementing a suitable actuarial formula or making a correct projection of mortality rates. You need very good and dedicated people.

Chapter 6 introduces what modern management experts might call the chief operating officer (COO) of the scheme—the colourful, controversial, but extremely efficient Reverend Alexander Webster. He was an integral part of the scheme from day one and was involved until his death 40 years later. Alexander Webster wasn't the only important personality who made an appearance, and other (more famous) individuals and members of the Scottish Enlightenment, such as the noted economist Adam Smith, Colin Maclaurin, as well as William Robertson, Francis Hutcheson, William Robertson, and Thomas Reid will make an appearance when I discuss the decisions they made to participate (or not) in the scheme for their widows and orphans.

Chapter 7, which is still in Part II of the book, then goes on to describe and analyse some of the more actuarial and financial aspects of the scheme, such as the asset allocation details, the actuarial projections, the valuation of the embedded annuities and the evolution of the fund over time. It is the only chapter with just a sprinkle of mathematical equations, and I'll delve into some of the (earliest) actuarial projections for an insurance fund. I'll also attempt to shed light on the fact that puzzled (and surprised) both Harari and Ferguson, namely, the accuracy of the projections for the value of the fund over time.

Moving on within Part II of the book, Chapter 8 goes back even further in time to the 1720s. It discusses how Presbyterian ministers of the Church of Scotland had attempted an earlier pay-as-you-go (versus funded) scheme that had failed to work because the "longevity risk sharing" pool wasn't set-up properly. These early ministers learned (the hard way) that pre-funding anticipated liabilities in advance is much more effective than passing a collection plate around, when funds are needed to pay benefits. So, if the 1740s was the genesis of pension auto-enrollment, then the 1720s was the dawn of the realization that pre-funding is better than PAYGO. Chapter 8 also discusses some of the controversies and crises within the Church of Scotland during that period, to shed some light on why the mid-1740s—when the funded scheme was launched—might have been the right time for such a venture and for more than purely scientific reasons.[22]

Chapter 9 then gets to the "religious roots" of the matter and examines the role of Protestant Christian theology—and in particular, the concept of predestination as understood by Calvinists and Presbyterians—on the probabilistic foundations of the scheme. There is a large and important literature on the development of probability during the eighteenth century, for example, *The Emergence of Probability* by Ian Hacking (2006) and the recently reprinted book *Classical Probability in the Enlightenment* by Lorraine Daston (2023). Neither of them discusses religious beliefs to any noticeable extent. But, in a time before the Law of Large Numbers was properly understood, I think that the architects of the 1744 annuity scheme truly believed the Lord would ensure that mortality rates were obeyed, and the requisite number of people would die at the appropriate times. This chapter of the book echoes and reinforces the argument, which is only noted briefly in Chapter 4, that religious thought, belief,

[22] See also Mepham (1988) and Lee (2021) for developments in Scottish accounting "theory and practice" during this time.

and practices have had a much greater influence on the development of annuity funding than previously acknowledged.

The idea at the core of this book is that religious thinking and institutions—and in particular the Presbyterian Church of Scotland—played a critical role in the development of best practices around the prudent (mathematical) risk management of pension annuities. The eighteenth-century founders of actuarial science and the trustees of *the* first funded longevity insurance scheme were devoutly religious individuals. These founders believed mortality rates obeyed "divine" probabilities and longevity modelling was just another branch of theology, thus giving them the confidence to make long-term financial projections that bordered on prophesies.

The last chapter of Part II, Chapter 10, concludes with a story of (possibly) the world's oldest (non-longevity pooled) life annuity, which is mentioned in the Hebrew Bible; quite apropos considering the religious theme of the book.

Then, **Part III** draws a continuous curve from the scheme in the mid-eighteenth century towards the future of longevity risk pooling in the twenty-first. The third part begins with Chapter 11, which is a detailed review and discussion of the many groups that opposed the establishment of such large (longevity pooling) plans and the rhetoric that was used to block their creation. Finally, Chapter 12 discusses the many challenges faced by pension pools in the twenty-first century. This last chapter wraps it all up and pulls it all together. Note that at the end of every chapter there is a self-contained list of cited sources, which also contains additional articles and readings that might be of interest.

Sources and Additional References

Bellhouse, D. R. (2011). *Abraham de Moivre: Setting the Stage for Classical Probability and Its Applications*. CRC Press.

Benjamin, D. J., Choi, J. J., & Fisher, G. (2016). Religious Identity and Economic Behavior. *Review of Economics and Statistics, 98*(4), 617–637.

Bruneau, O. (2020). Colin Maclaurin (1698–1746): A Newtonian Between Theory and Practice. *British Journal for the History of Mathematics, 35*(1), 52–62.

Buchan, J. (2003). *Capital of the Mind: How Edinburg Changed the World*. John Murray Publishers.

Daston, L. (2023). *Classical Probability in the Enlightenment* (New Ed.). Princeton University Press.

Deuchar, D. (1896). Notes on Widows' Funds. *Transactions of the Actuarial Society of Edinburgh, 3*, 198–227.
Dinkel, R. M. (1944). Attitudes of Children Toward Supporting Aged Parents. *American Sociological Review, 9*(4), 370–379.
Dow, J. B. (1975). Early Actuarial Work in 18th Century Scotland. *Transactions of the Faculty of Actuaries, 33*(240), 193–229.
Dunlop, A. I. (Ed.). (1992). *The Scottish Ministers' Widows' Fund, 1743–1993*. Hyperion Books.
Ferguson, N. (2008). Chapter 4: The Return of Risk. In *The Ascent of Money: A Financial History of the World*. Penguin.
Friedman, B. M. (2021). *Religion and the Rise of Capitalism*. Knopf.
Great Britain. (1744). *An Act for Raising and Establishing a Fund, for a Provision for the Widows and Children of the Ministers of the Church of Scotland, and of the Heads, Principals, and Masters, of the Universities of St. Andrews, Glasgow, and Edinburgh*.
Grünewald, A. (2021). From Benefits and Beneficiaries: The Historical Origins of Old-Age Pensions from a Political Regime Perspective. *Comparative Political Studies, 54*(8), 1424–1458.
Hacking, I. (2006). *The Emergence of Probability: A Philosophical Study of Early Ideas about Probability, Induction and Statistical Inference* (2nd ed.). Cambridge University Press.
Hald, A. (2003). *A History of Statistics and Probability and Their Application Before 1750*. Wiley.
Harari, Y. N. (2014). Chapter 14: The Discovery of Ignorance. In *Sapiens: A Brief History of Humankind*. Harvill Secker, Penguin.
Hare, J. P., & W. F. Scott. (1992). 'The Scottish Ministers' Widows' Fund of 1744'. In A. Ian Dunlop (Ed.), *The Scottish Ministers' Widows' Fund 1743–1993* (pp. 23–55). Saint Andrew Press.
Kumar, A., Page, J. K., & Spalt, O. G. (2011). Religious Beliefs, Gambling Attitudes, and Financial Market Outcomes. *Journal of Financial Economics, 102*(3), 671–708.
Lee, T. A. (2021). Edinburgh Accountants in Public Practice Pre-collective Organisation: 1757–1834. *Accounting History Review, 31*(2), 165–191.
LeFanu, W. R. (1921). *Queen Anne's Bounty: A Short Account of Its History and Work*. Macmillan.
Lewin, C. G. (2003). *Pensions and Insurance Before 1800: A Social History*. Tuckwell Press.
Mackie, A. (1956). *Facile Princeps: The Story of the Beginning of Life Insurance in America*. Presbyterian Ministers' Fund.
Mepham, M. J. (1988). The Scottish Enlightenment and the Development of Accounting. *The Accounting Historians Journal, 15*(2), 151–176.

Norris, C. M. (2017). Chapter 3: The Preacher's Fund. In *The Financing of John Wesley's Methodism C. 1740–1800*. Oxford University Press.

Perman, R. (2019). *The Rise and Fall of the City of Money: A Financial History of Edinburgh*. Birlinn Ltd.

Renneboog, L., & Spaenjers, C. (2012). Religion, Economic Attitudes, and Household Finance. *Oxford Economic Papers, 64*(1), 103–127.

Ryrie, A. (2006). *The Origins of the Scottish Reformation*. Manchester University Press.

Sher, R. B. (2015). *Church and University in the Scottish Enlightenment* (2nd ed.). Edinburgh University Press.

CHAPTER 2

Longevity Risk and Religion

Reading the Obituaries

This book revolves around the history of longevity risk pooling, with particular emphasis on *how and why* religious clerics, organizations, and intellectuals in the eighteenth century learned to manage that unique risk and were able to promise (better, more reliable) pension annuities. In order to take that journey, then, the first order of business is to gain a deeper understanding of the statistical nature of longevity risk, and in particular offer a suggestion as to why religious groups might care more about this longevity risk compared to other groups.

A few years ago, researchers in the American State of Ohio embarked on a project to investigate how long people live and hidden factors that affect longevity, using a rather novel set of data and facts: the obituaries section of their local newspaper.[1] Yes, that newspaper section might seem like an obvious place to look for reports of life and death, but it isn't the usual approach preferred by data scientists and statisticians in the twenty-first century. Those researchers and that field of inquiry tend to focus on official death certificates and government-collected mortality rates versus reports in the media, which could be biased and skewed. Nevertheless, these researchers from the Buckeye State—as Ohio is nicknamed—read, reviewed, and analysed over 1500 (randomized) obituaries of people who

[1] The article is cited in the bibliography as Wallace et al. (2019).

M. A. Milevsky, *The Religious Roots of Longevity Risk Sharing*, https://doi.org/10.1007/978-3-031-62403-2_2

had died during the year 2012 across 42 cities in the United States, and in particular they tallied or added up the average number of years lived by this group. Unsurprisingly and consistent with larger (national) databases and mortality statistics, the average number of years lived by these (1500 or so) random people was close to 76.5 years lived, with a *standard deviation*—to be explained—of nearly 16 years.

Again, the first average number is perfectly consistent with often-cited (period) life expectancy numbers in the United States, although it declined in the last few years as a result of the COVID-19 pandemic and is currently closer to 76 years. The second number reported—the *standard deviation* of 16 years—might not be as familiar to non-technical readers, but it is a measure of dispersion or uncertainty associated with lifetimes. It's an important number and at the heart and soul of managing longevity risk. Indeed, I'll revisit and use this measure of dispersion quite a few times throughout this book, and in particular in Chapter 7 and in the concluding discussion in Chapter 12, due to its critical role in monitoring longevity risk. So, here is a quick overview of standard deviation.

Intuitively, if you sample a large number of obituaries, some of the deceased might have been fortunate enough to live much longer than 76.5 years, perhaps beating the averages and reaching age 100 to become a centenarian—and others might have barely made it to the retirement age of 65. But, and this is the essence of the standard deviation calculation, if you re-examine the entire sample and compute the average distance of all these ages at the time of death from the (average) number 76.5, that number was 16.5 years, in the above-referenced study of the obituaries. Think of it as the *average* distance from the *average* number. To get an intuitive sense of this *standard deviation*, if everyone in the sample died at the exact age of 76, the average distance from the average would be zero, and so would the standard deviation estimate. Here is another very basic numerical example. If three people die at the ages of 60, 75, and 90, respectively, the average ages at death would be $(60 + 75 + 90)/3 = 75$, and the average distance from the average would be $(15 + 0 + 15)/3 = 10$ years.[2] The point is that the wider and broader the spread of lifetimes, the larger the standard deviation number. Generally speaking, you can

[2] In this example, technically, the 10 years would be the *sample* standard deviation, not the *population* standard deviation, and to be very precise (or I might lose my tenure in the mathematics department), the proper description would be the **square root** of the average **squared distance** from the average.

expect to see numbers in the order of 10–20 years for this dispersion or spread, when the underlying data is the age at death.

OK, now let's get back to the obituaries. Based on these 1500 life stories, achievements, and records of the deceased, the researchers found that females outlived males (on average) and that married people outlived single people (on average). Neither of these results or insights is surprising, and they are well-known to actuaries; I'll arrive at the truly and relevantly interesting result in a moment. But, to introduce yet another statistical term and measure, the researchers (obituary analysers) found that the correlation between longevity of life in years and marital status (a binary zero one variable) was between +30% and +40%. The number might be interesting but in and of itself the qualitative statement isn't very novel or path-breaking. In fact, many others (before and after them) have confirmed the beneficial effects of marriage on longevity, although it tends to be higher for males versus females.[3] To this point so far, the researchers hadn't uncovered or discovered anything new that hadn't been known to the hordes of statisticians, demographers, and actuaries who have studied mortality and longevity patterns for the last 100 years.

Now, finally, here is where the obituary study, which was eventually published in 2019 in the journal *Social Psychology and Personality Science*, got interesting, novel, and somewhat controversial. In fact, this is precisely the reason I'm beginning the journey to understand longevity risk and religion with this study. After computing the statistical averages, deviations, and correlations whose methodology was explained above, the researchers decided to split their dataset and the obituaries into two very distinct groups. The first group consisted of deceased people in which the obituary—which is more than just a death notice since it describes life achievements and milestones and community information—listed some sort of ***religious affiliation*** or participation in a religious organization. So, the obituary might have noted that the deceased was a member of a local church, synagogue, or mosque, or perhaps was active in a national religious organization, or perhaps that donations in memoriam should be

[3] See, for example, the article cited as Gardner and Oswald (2004), where the examine how mortality is affected by money, marriage, and stress, and was published in the *Journal of Health Economics*. They write: "Marriage is found to be associated with substantially lower rates of mortality, for both men and women. Married men are predicted to be some −7.2 per cent less likely to die over the period than unmarried men. For women, the effect is smaller."

sent to a religious group that was a favourite of the deceased. These were considered religious people. Yes, there is quite a bit of subjectivity (and work) involved when interpreting these descriptions, and this isn't the only study I'll be discussing in the chapter. The subjectivity involved and the time it takes might be two of the reasons the study was limited in data size and scope.

To the main idea here, those obituaries that didn't list any religious affiliation, organization, or connection whatsoever were classified as being *non-religious* individuals. Hold on and bear with me here for a moment. In total, between 40 and 50% of the obituaries examined by the researchers listed some sort of religious affiliation, and the lives and their longevity (in years) were coded as religious. The remainder of the obituaries—the precise methodology used goes beyond the scope of this chapter—were coded as non-religious lives.

One thing is important to note about the statistical methodology, and that is that each obituary—and the number of years the deceased had lived—was either classified as a zero (non-religious) or a one (religious). There were no fractions or partial marks allocated to religiosity. This binary nature of the religious classification of obituaries and the complicated people they attempted to represent is a weakness of the study, and some might consider a flaw in any results. Nevertheless, to remind us of this fact, from now on I will make statements about religious obituaries and non-religious obituaries, vs. people. Indeed, you might already sense why the study itself was controversial. Though, allow me to get to the conclusion—which you can intuit from the very long windup—before I return to these and other methodological matters. Which group of obituaries do you think lived longer on average? Was it the religious or the non-religious?

It turns out that religion and longevity go together, but only on average. To begin to think about and explore that relationship, and understand longevity risk in general, Fig. 2.1 nearby[4] offers a visual but abstract picture of these two groups. To use the actual study's authors' words: "In two samples the religiously affiliated lived 9.45 years to 5.64 years longer, respectively, than the non-religiously affiliated."[5] With those words in mind, here is how to interpret the figure, which again is an abstraction.

[4] Created and generated by the author.

[5] Again, the study is cited as: Wallace et al. (2019).

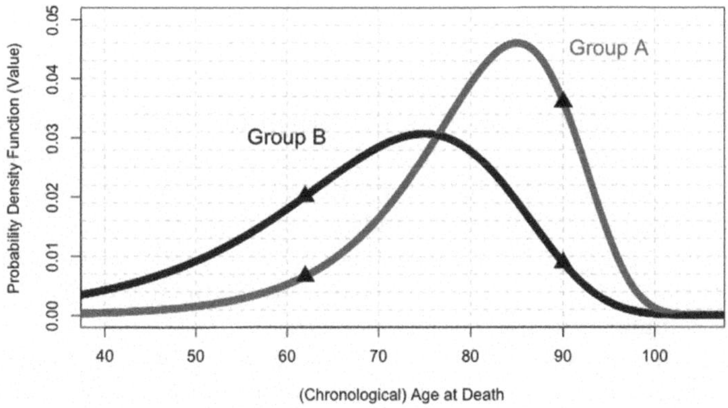

Fig. 2.1 Visualizing longevity risk and dispersion

Think of Group A (in red, for those who can see this in colour) as consisting of religious obituaries and Group B (in blue) as consisting of non-religious obituaries. Each point on the two curves—which statisticians label a probability density function—represents one of the obituaries. Group A obituaries cluster to the right and live longer than Group B. On average, members of Group A live longer than members of Group B by almost 12 years. In the figure, the life expectancy of the group (A vs. B) is *approximately* the point on the curves where the peak is the highest. Or stated even more loosely, the peak of the red mountain (Group A) is about 12 years to the east of the peak of the blue mountain (Group B.) Nevertheless, and this is very pertinent to the word "risk" in the term longevity risk, some members of Group A (e.g., the religious obituaries) only live to ages 40, 50, or 60, and some members of Group B (e.g., the non-religious obituaries) live into their late 80s and 90s. That's the nature of longevity risk. In some sense, the dispersion around the peak (of the mountain) is more important than where the peak itself is located. Longevity risk is not associated with the height of the peak but the dispersion around the peak.

I must continue to dig a little bit deeper into the statistics behind Fig. 2.1, since it has an important and recurring role in the description and management of longevity risk.[6] To recap, the weighted average age of death for members of Group A is approximately 68 years while for members of Group B it's 80. Again, this gap of 12 years is why the Group A (in red) curve is to the right of the Group B (blue) curve but notice carefully that Group A is more peaked, clustered, and tighter around the modal (i.e., most likely) value. What this means mathematically is that the above-noted and explained standard deviation of Group A is lower compared to Group B. For those readers who might be familiar with MBA portfolio theory, Group A stochastically dominates Group B, since the mean is higher and the variance is lower.

Beyond the shapes and statistics, there is additional information contained in the figure that can help provide intuition for longevity risk. Focus on the (black) triangles placed on the curves at the ages of 62 and 90. Here is how to interpret those marks. A member of Group B is three times more likely to die at the age of 62 compared to a member of Group A. The triangle mark on the blue curve is at a height (or value) that's three times greater than on the red curve. If you are on the other side of these curves, if you are a member of Group A, then you are four times more likely to die at the age of 90 versus those in Group B. You can read across to the y-axis values to see the relative values.

To state the obvious, everyone dies in the end or at least by the age of 122 (which is the current record), but it's just a question of where those dates are spread across the horizontal age axis. For Group A, this event happens later than for Group B, but nothing is guaranteed other than eventual death (and taxes).

The one final way to state this is by comparing the values of the triangles themselves, on the same curve. If you are in Group A, you are 5.4 times more likely to die (old) at 90 versus (young) at 62. But if you are in Group B, you are 2.3 times more likely to die at (young) 62 versus at (old) 90. Those are the relative probabilities. My lecture on interpreting probability density functions is complete, and I now return to the study of obituaries, religious affiliation, and their longevity edge.

Remember that although Fig. 2.1 is an abstract representation of the study results, and one that I created myself, my point is that Group A

[6] Technically, the Group A density function, conditional on age zero, was generated using a Gompertz mortality model with parameters (88, 8), and Group B was generated using parameters (75, 12). See Milevsky (2020), Chapters 7 and 8 for more on the Gompertz model.

has a religious affiliation and Group B doesn't. The practical takeaway from the study published in *Social Psychology and Personality Science* is as follows: Something prolongs life for the religious, while the life of the non-religious is shorter.

This empirical result—which hasn't gone unchallenged—might seem fabulous and God-sent for those who have a religious affiliation, and somewhat disturbing for those who might code themselves as having zero religious affiliation. Some might wonder: Is this a one-off and non-replicated result that hasn't withstood the test of time? Has it been retracted? Well, the authors themselves take pains to emphasize that this isn't a new discovery in the social sciences. Rather, it's their methodology of reviewing obituaries that was novel. And, even after controlling for a number of confounding factors, the authors of the study conclude their research paper by writing: "In conclusion, we use a novel methodology to corroborate survey data associating religious affiliation with longevity."[7] They are corroborating and confirming something social scientists have long known, they did not discover anything new. Since the authors themselves took pains to emphasize to use the word "corroborate" versus the word "discover," and I'll mention some other studies in the next few pages.

Before I survey some of the other (supporting) studies of the same idea, let me be clear about something these (obituary reading) authors never said. In fact, the article never even hinted, let alone declared triumphantly, that people with religious beliefs and those who believe in God live longer than those who don't. They never wrote that, and I'm not saying it either. Remember that they were examining obituaries—and not speaking to live people—which is why they were careful to use the phrase *religious affiliation*. They found that those with an attachment to, or membership in, a religious group ended up living longer than those who did not, and only on average.

Nevertheless, this study and others along the same vein have been subjected to quite a bit of methodological criticism. The criticism isn't about channelling Voltaire against organized religion—who coincidently lived to the ripe old age of 83 back in the eighteenth century—rather,

[7] Wallace et al. (2019, p. 668).

the challenges are entirely statistical.[8] The statistical problem is that in addition to the (rather implausible) zero–one nature of their analysis, religious affiliation in the twenty-first century might be a proxy for something completely different. What is that different thing? Sociologists and psychologists might say: The other different thing is the repeated and systematic participation in activities of a *social group*, perhaps being part of a large cohesive pool of similarly minded people. Before I delve into what religious affiliation might really be capturing, here's yet another study that sheds light on the matter.

Visiting the Gravestones

Not to be outdone by obituaries, a separate group of researchers decided to examine the same question: *Is religious affiliation associated with greater longevity?*, but used a very different type of dataset. Instead of scouring newspapers, they visited cemeteries to look at gravestones.[9] Recall that obituaries tend to be quite lengthy and lack any natural constraint on the amount of text and verbiage they contain—other than the marginal cost of an additional word—so the reader never knows how much is exaggerated or added by surviving family members. On the actual gravestones, though, the deceased or their loved ones must pick words and *symbols* wisely. Space is at a dear premium. Much more thought goes into each inch of space. With that constraint in mind, the researchers examined the inscriptions and imagery on a total of 6,400 gravestones across 64 urban and rural counties in the United States. They were looking for crosses, Bibles, praying hands, scripture, and angels, all of which might indicate the religiosity of the person buried beneath. Similar to the study on the obituaries, the researchers tallied the total number of years the deceased had lived—information that was usually available from the gravestones themselves since dates of birth and death are noted—and compared the count against those buried in the same cemetery whose gravestones were absent of any religious imagery.

[8] Voltaire was a noted critic of the church and religious beliefs, who no doubt would have chafed at the (erroneous) statement that belief in God keeps someone alive longer.

[9] The formal article is cited as Ebert et al. (2020), with the self-explanatory title: religious people only live longer in religious cultural contexts: A gravestone analysis. It was published in the *Journal of Personality and Social Psychology*.

Methodologically, they randomly sampled 100 deceased from each of the 64 cemeteries—of which 50% were males and 50% were females—all of whom passed away after January 1, 2000. For the sake of simplicity—which might indicate a slight religious bias—the gravestone researchers limited themselves to Christians, so the Star of David or the Muslim Star and Crescent weren't included in the samples. And, while all this wandering around graves might sound like an untowardly and macabre amount of time to spend in cemeteries, this project wasn't as gruesome as it sounds. In fact, the researchers used the website www.findagrave.com to conduct their search. The website contains 226 million entries, so the team likely did this from the comfort of their home, office, or local coffee shop.

As in the zero–one binary indicator for religious affiliation used for the obituaries, some readers might be (even more) sceptical that such iconography truly represents or captures whether the individual buried under the gravestone was religiously affiliated. For this reason, the authors also took great pains to dig into the lives of some of the deceased buried under those same gravestones to eliminate other possible explanations for the religious iconography. In defending this visual methodology, the team of authors, who were from notable universities in Germany, Denmark, and the UK, wrote: "Taken together, evidence from the aggregate level and the individual level converged, suggesting that the deceased person's religiosity can be validly assessed by the presence of books, praying hands and angels on that person's gravestone." Having justified their approach, the authors then reported that 44% of the gravestones sampled included religious imagery, which coincidently isn't far from the nearly half of obituaries in the prior study that were classified as having a religious affiliation.[10]

Having been primed by the results of study of obituaries, the results of the gravestone analysis shouldn't be surprising. Gravestones told the same tale, namely, that those with religious imagery lived longer than those without the imagery. Having crosses, scriptures, or angels on your own gravestone is indeed associated with a longer life. The abstract Fig. 2.1 is a good intuitive indication of the impact of having religious iconography on your gravestone, although remember that it's too late to add them after you have died! However, as far as the actual magnitude or the number

[10] Ebert et al. (2020, p. 3).

of years gained, the results of the study were more subtle and rather nuanced compared to the obituaries. The average gap between members of—what I have abstractly called—Group A and Group B depended on the particular cemetery in which they were buried. More precisely, the average longevity gap depended on the religiosity of the county within which these 64 cemeteries were located. This is what the authors refer to as the religious cultural context, which I'll explain.

In counties that were classified as religious—which was limited to the county having a high share of Christians—the average age at death for those with religious imagery on their gravestones was 86.8 years, versus an average age at death of 84.6 for the remainder. In those counties, the crosses, scriptures, and angels added an average of 2.2 more years of life. This is a number or longevity benefit that is much smaller, compared to the 5–10 years from the earlier obituary study. And, here is an even more sobering result for those hoping to capitalize on religiously induced longevity. In the counties with the lowest levels of religiosity, the most religious graves showed that individuals lived 1.7 years less (yes, a deficit) than individuals in the non-religious graves. Again, this is all based on gravestone iconography, and none (obviously) of the individuals were interviewed about their spiritual beliefs or religious attendance. This is why the authors titled their paper "Religious People Only Live Longer in Religious Cultural Contexts." Notice the disclaimer about culture and context.

Incidentally, and certainly not the main focus of their analysis, the authors of the gravestone study noted something interesting and quite relevant to the topic of pensions and annuities. They commented that religion—very generally speaking—only begins to show up in the mortality data after the age of 70 and that the greatest difference between religious and non-religious longevity patterns occurs between the ages of 90 and 100. I'll get back to whether religious affiliation is more effective with age in the next subsection. For now, though, I will summarize the gravestone article with a pertinent quote from the authors.

The authors conclude and write: "Our study holds important implications for a Western World that is becoming increasingly secularized. Religiosity does not delay death everywhere and rising secularization will probably not curb collective life expectancy in the future."[11] Overall, one gets a more nuanced and cautious image of how religion affects longevity.

But neither obituaries nor tombstones have provided the final word on this matter, and the hunt for the secrets to longevity continues. Other researchers, with larger datasets, more powerful econometric techniques, and perhaps less subjectivity, have doubled down on the benefits of religious affiliation and longevity. But, before I delve into the proposed explanations for why this affiliation is beneficial, I'll share the results of one final study that—though it might not be as interesting as obituaries or gravestones—is much more convincing, partially because the authors are such authorities in the field.

Deaths of Despair

In the year 2020, two professors at Princeton University—Anne Case and her husband, Nobel Prize–winning economist Angus Deaton—published a rather sobering book that you might have (and probably should) read called *Deaths of Despair*.[12] This book, which is based on a series of published articles by the authors, documents a type of mortality effect or cause they discovered and coined. Namely, while US population mortality rates have declined and longevity has increased in the country during the last century, over the last few decades this positive and improving trend has broken down among white middle-aged males. Sadly, certain death rates have been increasing steadily, something which can't be blamed on the COVID-19 pandemic as this began as far back as the late 1990s.

According to Case and Deaton, the death rates—from what the authors christened *despair* among white American individuals between the ages of 45 and 64—have soared in the last few decades. The phrase "deaths of despair" is used because drug poisoning deaths have tripled, deaths by suicide have increased by 30%, as have deaths from alcoholic liver diseases. With their book, Professors Case and Deaton unleashed a torrent of academic interest on the topic, partially as a result of their high-profile media attention. So, an open research puzzle or question in US social

[11] Ebert et al. (2020, p. 5).

[12] The book was written by Anne Case and her husband Angus Deaton, published in 2020, and received with much publicity. It is called: *Deaths of Despair and the Future of Capitalism*. See also the 7 December 2023 article in the *Economist Magazine*, entitled "At last, a convincing explanation for America's drug-death crisis."

science circles is to understand why these deaths of despair have occurred and what might be done to reverse this trend.

A culprit that has been suggested for why white males between the ages of 45 and 64 are dying at a disproportionate and high rate is OxyContin, a drug marketed by Purdue Pharma. But these high rates of death appear to have started in the late 1980s, well before OxyContin, which was introduced in the year 1996, so the search continues for alternative explanations.[13] To put this in a graphical context, think of the Group B curve in the abstract curves within Fig. 2.1, and its shifting to the left (the mountain moves west) instead of the right (moving east) as time goes on. Although to be very clear, I am not discussing all-cause mortality for the entire US population, but rather the deaths from three very specific causes and for a limited group. Also, this is limited to white males. Deaths of despair aren't being observed among retired Latinos, Asian, and African Americans, males or females. Nevertheless, answering the question of *why* this particular group is suffering from excess mortality—to use another popular term—is on many bright minds.

Well, getting back to religious affiliation, one possible explanation for these deaths of despair, and one that might resonate with readers in light of the earlier studies, is the overall decline in religious participation among middle-aged white Americans during the last few decades. Less religion leads to more despair and then eventually death. Yes, this is a gross oversimplification to theorize that less religion leads to higher rates of mortality, but don't take my word for it. Noted economists and researchers affiliated with the prestigious National Bureau of Economic Research (NBER) in the United States have collected some unique data to help prove this link. The NBER researchers begin their research article by writing that, in the United States, the "states that experienced larger declines in religious participation in the last 15 years of the twentieth century saw larger increases in deaths of despair." Moreover, the decline in religiosity that they observed and the rise in the deaths of despair were driven by similar groups and were in similar locations. States with "high levels of religiosity have suffered less from mortality due to alcohol, suicides and drug poisonings." It also appears that after the US Supreme Court repealed or greatly reduced the scope of the blue laws,

[13] The main reference for this paragraph and the next is the working paper cited as Giles et al. (2023).

also known as the Sabbath closing laws, and thus allowed for the expansion of commerce on Sunday mornings, religious participation declined and deaths of despair increased. To these economists and statisticians, the data all point in the same direction, and the thesis I offered earlier. Namely, mortality rates and religiosity rates—notwithstanding the difficulty in measuring the latter—are negatively correlated. Stated differently, longevity and religiosity are positively correlated.[14]

Social Versus Religious Components

I am now ready to address the bigger question: *What is it about religious affiliation or participation or even beliefs that reduces mortality and thus increases longevity?* To begin, there is one possible hypothesis that was alluded to in quotes from the Book of Proverbs at the very beginning of this book: an omnipotent deity is rewarding his (or her) followers with a longer life or, for those who are more statistically minded, a left shift in the curve of Fig. 2.1. Many of the eighteenth-century Protestant Christian ministers who dabbled in early probability theory, and who will be introduced in Part II of this book, would no doubt endorse this divine theory. To these ministers, the truth was simple, if you believe in God then he (usually male) would keep you alive longer. If you didn't, he would kill you sooner. Period.

Now, in contrast to pinning this on a deity, more pragmatic researchers—many of whom are sociologists—have delved into the precise nature, activities, and rituals of religious affiliation that might promote longevity. In fact, the authors of the above-cited obituary and gravestone studies all alluded to the simple fact that participation in *social and group activities*, regardless of any religious nature, can have beneficial health effects. Why? Well, recreational group activities such as social clubs and voluntary associations create a sense of acceptance, inclusion, and companionship. These activities can help reduce loneliness and social isolation, which is a noted and important factor for increased mortality. Being part of an organized social group might also intervene to reduce unhealthy behaviour and provide normative and moral guidance. The act of volunteering generates a sense of purpose and offers an individual a specific role to play, which then cascades into physical and mental health

[14] Giles et al. (2023, pp. 3 and 30), see also Idler et al. (2017), McCullough et al. (2009) and Pargament et al. (2001), for more on longevity and religion.

effects that can reduce mortality. These are all alternative explanations that are unrelated to religiosity per se, or to spirituality, let alone the act of a deity. And, to many scientists, these explanations are much more palatable than God keeping them alive longer.

A meta-analysis, which is a scientific term for an aggregation of 74 different studies on religiosity and mortality, made these same arguments. A summary of this collection was published in the year 2013 in the *Journal for the Scientific Study of Religion*. It set the stage for the main hypothesis by noting that "the association between high levels of religious participation and reduced mortality risk is often thought to derive from a social participation component and from a religious efficacy component." But, after reviewing these 74 different studies, the authors concluded that "the analysis reported here, however, does not support the religious efficacy hypothesis."[15] Rather, as alluded to just above, it is the low level of participation in social activities that can increase mortality rates. The consensus is that religion itself, the beliefs, the spirituality, and the deities have little to do with it.

To the researchers, the protective health and mortality benefits observed in the data—which they do not dispute or question—come from the increase in social support as well as a sense of self-worth that they provide. To use the language of statistics, they find no significant differences between the "mean hazard ratio" (i.e., the mortality rate) of groups with low religious participation (i.e., non-religious) and groups with low non-religious participation (i.e., religious). Back again to Fig. 2.1, the difference between Group A and Group B isn't what the group members believe or whether they participate in their local church, synagogue, or mosque regularly. Rather, it's all about whether they participate in *any sort of activity* regularly. Here is practical advice for 21st century longevity. Retirees, get out, get active and ideally do it together with other like-minded people. This isn't news. But in the 18th century, the belief would have been that it's all about belief.

The French sociologist Emile Durkheim (b. 1858, d. 1917)—well known for initiating the systematic study of primitive religions within society—alluded to the idea that participating in the rituals of religious worship had some benefits. In his words, "Believers…feel that the true function of religion is not to make us think…but to make us act, to

[15] The article is cited as Shor and Roelfs (2013).

help us live. He feels more strength in himself, either to cope with the difficulties of existence or to defeat them."[16]

In sum, you may have your own personal views on why the longevity odds are better for those in the religiously affiliated Group A versus those in the non-religiously affiliated Group B, and no doubt the debates around reasons will continue. Moreover, the odds favour a future scenario in which journals and outlets such as the *Journal for the Scientific Study of Religion* continue to publish studies on the topic. In fact, with the growing availability of what's called "big data" and artificial intelligence (AI) techniques that can mine and interpret (and perhaps torture) growing troves of survival data, I suspect there will be many more researchers joining the fray and trying to explain *if* religious affiliation is indeed special and *why* it's associated with greater longevity. But the existence of a noticeable longevity effect—measured in years—within religious communities might soon be affirmed as reliable as Isaac Newton's law of gravity, a man who, for the record, was quite religious and died at the (above-average) age of 84.

Clergy Versus the Rest

If one observes that religious affiliation positively affects—or at the very least is associated with more—longevity, regardless of the precise channel by which this reduces mortality, one might also expect to see this in mortality data for members of the clergy. These are the front-line managers of the religious organizations, clubs, and groups. Indeed, this was precisely the finding and results published by a group of US researchers in a 2002 issue of the *Journal of Religion and Health*.[17] The clergy live longer too.

For background and context, at the turn of the twenty-first century—when the study was conducted—there were close to 400,000 Christian and Jewish clergy in the United States, of which 4000 were Jewish rabbis, 49,000 were Catholic priests, and 300,000—yes, a much larger group—were Protestant ministers. The authors didn't mention Muslim imams or clergy from other faiths, which is a religious group that is (sadly, mostly) absent from numerous longevity studies. Nevertheless, the authors of the

[16] Durkheim (2001).

[17] Flannelly et al. (2002).

study in the *Journal of Religion and Health* conducted a meta-analysis, or more precisely a comprehensive literature review of all articles written during the prior 40 years that examined the mortality rates of Jewish and Christian clergy relative to the rest of the population.

The authors found that, for the most part, prior studies had supported the hypothesis that clergy experienced lower death rates from all causes compared to the rest of the population. To be precise, of the prior studies reviewed, 11 of the published papers had found that clergy outlived everyone else, and only one lone study was inconclusive. In these 12 studies that were summarized by the authors, the standardized mortality rates of clergy were between 68% (i.e., much lower) and 103% (roughly the same) of the corresponding population mortality rates. In other words, the clergy had a lower mortality rate and thus lived longer, other than that one sole study that reported a slightly elevated (103%) risk of death. In the authors' words: "Compared to American males, the clergymen were found to have an advantage for almost every cause-specific mortality rate that was examined." On a related note, when one considered causes of death, the authors of the study—which was published in the *Journal of Religion and Health*—were careful to note that mortality from leukaemia, lymphoma, and hypertension was similar in the general population. So, the reduced mortality rates were not observed across the board from all causes of death but were limited to certain specific diseases.

Now, by this point in the narrative, I suspect that the lower mortality rate for religious clergy shouldn't surprise you; in fact, it is perfectly consistent with the results applying to those who are religiously affiliated. But, here is the final result I'm driving towards and what was (even more) unexpected, the *Journal of Religion and Health* authors found that standardized mortality rates for Catholic Christians versus Protestant Christians were higher. In their words: "Mortality rates among the US clergy were the lowest among the Protestant denominations." According to their calculations, ministers from a Protestant Christian denomination had an average 25% mortality advantage compared to the general population. I should note that this so-called advantaged Protestant group is quite broad, and includes clergy from the Baptist, Lutheran, Protestant, Episcopalian, and United Presbyterian Church in the United States, whereas Catholic priest or (even) Jewish rabbi is a much more homogenous category.

Nevertheless, the Protestant versus Catholic result is unusual but rhymes with the pension ranking data in the first Chapter 1. Having a religious affiliation is associated with greater longevity, and Protestants (clergy) are blessed with longer years.[18]

Longevity: Now Versus Then

Now, the studies and data presented over the last few pages have been exclusively focused on the period between the mid-twentieth century and early twenty-first century. Considering the historical focus of this book, readers might rightfully wonder if these mortality and longevity advantages for the religiously affiliated were observed in earlier centuries. Might this all be a recent phenomenon? Perhaps it's due to modern medicine, sanitation, and health practices? Well, it seems the answer is no, it's not a modern phenomenon. According to a study published in the *Journal of Biosocial Science*, these religious (affiliation) differences were observed as far back as the eighteenth century.[19]

The historically oriented researchers of *Journal of Biosocial Science* study obtained Parish records of births and deaths from the Austrian village of Hallstatt, covering the years 1733 to 1908. That particular location and city might seem like an odd, completely random, and very unlikely place to investigate longevity, but it is actually a demographic and religious laboratory. Here's why Hallstatt deserves your attention and is quite pertinent to the question of longevity for Protestants versus Catholics versus the rest.

During the Protestant Reformation, the city of Hallstatt and its community was effectively split down the middle into two parts, one Catholic and the other Protestant. Parishioners in the latter group created their own new records and registries of births and deaths, while the former group continued compiling these life events as they had done in the past. The results were two distinct sets of parallel records over two centuries for the same city. This natural laboratory allowed the researchers to create separate Catholic and Protestant data records, genealogies, and life histories from a group that had been nearly identical and lived in the same geographic location.

In addition to information about dates of marriage and number of offspring, the researchers compiled ages at death for 2,351 Catholics and

[18] Flannelly et al. (2002, p. 63).

[19] Gavrus-Ion et al. (2021).

for 1,334 Protestants from Hallstatt who had survived into adulthood. A small sample to be certain, even when compared with the obituaries and tombstone studies, but a sample large enough for statistical significance.

The results were that the median age of death for Protestants was 64.2 years, while the median age of death for Catholics was 58.8 years. This was a 5.4-year longevity advantage *on average*. Again, it wasn't guaranteed or a certainty for every Protestant Christian. In fact, the authors were careful to compute the range of lifespans—think bach to Groups A and B in Fig. 2.1—and reported that for Catholics the interquartile range was (38, 72) years and for Protestants it was (46, 76) years. Notice the slight shift in range to the right (longer life) for Protestants. Overall, the median numbers themselves might seem low for both groups, but remember these averages are during the eighteenth and nineteenth centuries. Yes, the life expectancy numbers are old and stale and are only separated from each other by religious affiliation. But that's exactly the point: they illustrate that these effects—Catholic versus Protestant longevity patterns—were alive and well at the dawn of annuity financing.[20]

Did People Know Their Risks?

It's unclear whether at the time (that is the 17th and 18th century) such data were being compiled, whether people knew or even understood longevity averages, let alone longevity risk. The statistical science behind longevity was just beginning to emerge—in ways I will explain in later chapters—and even the concept of "life expectancy" was undefined. We can all look back from our civilized perch in the twenty-first century and declare that religiously affiliated groups have a longer life expectancy, or that Protestants (ministers) might live longer than Catholics (priests), but that sort of statement is anachronistic. It's unlikely they knew this, then, because the concept and measure itself didn't exist.

In fact, there is empirical evidence to suggest that even in the twenty-first century many consumers fail to appreciate the full statistical magnitude of longevity risk and the odds of living to very advanced ages. That is the right-hand "tail" of the central Fig. 2.1. In laboratory experiments and corresponding surveys, when random respondents are

[20] Source for numbers in paragraph is Table 1 in Gavrus-Ion et al. (2021, p. 309).

asked "So, how long do you think you will live?" the replies range from unrealistic to entirely ridiculous.

A particularly puzzling finding was unearthed when a team of academic researchers asked a large group of soon-to-be-retirees to estimate the probability they would be *alive* by the age of 85. The question was coached in optimistic language: alive. This elicited average numbers around 55%, which in and of itself wasn't problematic or surprising. People thought the odds they would be alive were close to even. But alas, when the exact same group was asked a follow-up question, namely, the probability that they would be *dead* by the age of 85, which gave the question a more negative connotation and language, the response was around 70%. Again, this was the exact same group. Now, I'll pause for a moment to let these two numbers sink in. Think about it. Live to 85, around 55%. Die by 85%, around 70%.

The response makes absolutely no sense and is inconsistent with the first question. Life and death probabilities must add up to 100%, no more and no less. If there is a 70% probability of being dead, there is only a 30% probability of being alive. So which one is it—30% or 55%? Indeed, the results of this particular study—pointing to deep innumeracy in addition to a language priming affect—were published in the prestigious *Journal of Risk and Uncertainty*. The authors concluded: "We find that individuals in a [optimistic] live to frame report significantly higher chances of being alive at age 55–95, than people in a corresponding [pessimistic] die-by frame. Longevity risk is not only in the eye of, but also the language, framing and words of, the beholder."[21]

In a related study by Wharton Professor Olivia Mitchell and her co-authors, they noted that "many people do not understand the concepts of life expectancy and longevity risk, potentially leading them to under-save for retirement or to not purchase longevity insurance."[22] As to why consumers might have such distorted perceptions—or plain ignorance of—their longevity (risk) is a subject of ongoing psychological research and an area that risks taking us far away from the historical roots.

I should also briefly mention a recent article that appeared in the *Journal of Consumer Research* that helps address why people might not

[21] Cited as Payne et al. (2013).

[22] See Hurwitz et al. (2022).

understand or misestimate the basic idea underlying Fig. 2.1. One explanation for longevity pessimism was what the authors coined as the effects of environmental stressors and, more specifically, childhood poverty. In their words: "People who grew up in poorer environments expect[ed] to have a shorter lifespan compared to people who grew up in richer environments." A surveyed group of wealthier respondents whose median age was 35 expected to live to age 85, but those from "poorer backgrounds expected to live significantly fewer years." This is another important factor in longevity risk. Expectations of life differ between groups, and their perception of those expectations does as well.[23]

A Religious Model for Longevity

In early 2023, the Organisation for Economic Co-operation and Development (OECD), headquartered in Paris, released a report titled *Mortality and the Provision of Retirement Income*. This 140-page document, intended for distribution to all 36 member countries of the OECD, provides technical guidance to government regulators and pension supervisors. The report examines the trends in mortality assumptions used in the provision of retirement income to shed light on different approaches taken to estimate mortality rates and inform future expectations of mortality trends.

In fact, if you are in—or even tangentially related to—the business of predicting mortality rates for the purpose of valuing pension benefits, reading the OECD report is recommended. Estimating or projecting how much income a plan can afford to pay future retirees depends critically on forward-looking mortality assumptions regarding how long those retirees are expected to live. The authors also emphasize the importance of the "historical drivers" of mortality, which helps inform future expectations.

The report is replete with technical tips on how to "model mortality," but it also details some concrete statistical facts that resonate and dovetail nicely with the insights from this chapter. Across OECD countries, life expectancy at age 65 ranges from a mere 13.6 years for Lithuanian males to 24.6 years for Japanese women. The global longevity dispersion range is rather wide. Since 1990, countries within the OECD have gained an average of between 5.7 and 7.4 years of life expectancy at birth, for males

[23] Mittal et al. (2020).

and females, respectively. More importantly, groups with a higher socio-economic status tend to have a higher life expectancy at the baseline age of 65; possibly 2.5 years of additional life, and this gap is widening over time. It is therefore not surprising that longevity risk is a hot topic.

The OECD report also notes (or warns) that aggregate mortality rates continue to decline around the world and therefore "static assumptions" that do not account for that ongoing improvement might understate future life expectancy by as much as 1.5 years. Stated simply: We might all live longer than we think.

The OECD authors note the recent stagnation in US death rates, which was mentioned earlier in this chapter, and the sharp increase in mortality related to liver disease, suicide, and drugs. Equally sobering, at the oldest of ages, dementia appears to be causing an increase in mortality that goes against the positive trend. Nevertheless, the authoritative OECD report repeatedly emphasizes that retired pensioners and annuitants are different from the rest of the population because of the higher incomes they receive, and therefore, general trends might not be applicable to this select group. In sum, the OECD report encourages policymakers to be "open to innovative approaches" when estimating and forecasting future mortality, even as they acknowledge the difficulty in modelling longevity risk, especially after the COVID-19 pandemic.

However, it's worth noting—and apropos for the conclusion of this chapter—something that is missing from the 140-page OECD report. Nowhere in the report is there a mention of religion, religiosity, or culture, or even of how social communities might affect mortality and longevity. While the OECD—with headquarters in Paris, France—is unlikely to promote or advocate or endorse religious institutions, it's noteworthy that even the phrase "cultural factors" isn't mentioned as a factor, anywhere.

In light of the studies summarized in this chapter and the larger universe of reports not mentioned in this brief review, it is really puzzling that religion is nowhere to be found in the OECD report.

This isn't a complaint about the OECD research, alone. In fact, a recent report by the World Economic Forum and Mercer consulting—that same Mercer organization from Chapter 1—published a report on *longevity literacy*, which discusses the factors and drivers of a long life.[24] Again, the words religion, faith, or beliefs do not appear anywhere in the 70-page report. I suspect that this is an omission that is symbolic of a broader trend that will echo the main idea of this book. Religion has not only been removed from the future of longevity innovations, it has also been cleansed from its past.[25]

Conclusion for a Long Life

Despite the length of this chapter and occasional digression, there are a number of important takeaways worth emphasizing and noting. They are all related to the book's central theme of longevity risk, which must begin with a detailed and careful understanding of what exactly is meant by a so-called risk that many would view as a blessing: longevity.

First, life expectancy is just the average summary number of a large collection of individual numbers, representing how long people actually live, from zero to (possibly) 122. The reasonable range of longevity outcomes can be quite wide regardless of what socio-economic or religious group you belong to and/or pray with. So, it's best to think in terms of curves and probabilities. When someone asks themselves (or someone else) how long they will live, the answer really is not a number (of years) but rather a curve (of years). The second takeaway is as follows: the correlation or statistical relationship between religious affiliation of an individual and their mortality rate is negative—which (again) is a positive outcome if you have such a communal attachment. Stated differently, individuals with some sort of religious affiliation or connection or involvement are expected to live longer. This is an expectation, an average, and a tendency and not a guarantee or a certainty. Think (again) curves and not numbers. Moreover, the exact magnitude of this longevity effect is unknown exactly, and while it depends on the cultural context, it can be measured in years.

[24] This report can be found at https://www.weforum.org/agenda/2023/01/longevity-literacy-secret-to-prosperous-longer-life-davos-2023/?DAG=3&gclid=CjwKCAjwyqWkBhBMEiwAp2yUFo14M5adouMXrecS-f2bGfuoC6Ti7LaKIUQj6UV0NaE7kKEGZadYbhoCzIIQAvD_BwE, and the report is titled: "Why longevity literacy is the secret to a prosperous longer life."

[25] On this note, see the 10 June 2023 article in the *Wall Street Journal* titled: "Is Religion Good for Your Health?" (https://www.wsj.com/articles/is-religion-good-for-your-health-921814a7) It echoes many of the arguments that appear in this chapter.

Third and finally, my intelectual leap of faith. If religious practices and affiliation and community can add two, five, or perhaps even a decade to the length of life—regardless of the reasons—it might make sense for a "religious organization" to take the lead in leveraging their financial resources to help finance this longer life. These organizations are most likely to have among them (and certainly notice) elderly individuals and might even feel a religious obligation to care for the elderly. I'm not saying anything (at this point) about widows and orphans, or Protestant versus Catholics. The only point I'm making here is that if a particular group experiences (or is blessed with) an abundance of longer years, they might notice and want to do something about it. That's it. This then brings me to the next topic and chapter, the (magical) mechanics of longevity risk pooling.

Sources and Additional References

Case, A., & Deaton, A. (2020). *Deaths of Despair and the Future of Capitalism*. Princeton University Press.

Durkheim, E. (2001). *The Elementary Forms of Religious Life* (C. Cosman, Trans., p. 311). Oxford World's Classics.

Ebert, T., Gebauer, J. E., Talman, J. R., & Rentfrow, P. J. (2020). Religious People only Live Longer in Religious Cultural Contexts: A Gravestone Analysis. *Journal of Personality and Social Psychology, 119*(1), 1–5.

Flannelly, K. J., Weaver, A. J., Larson, D. B., & Koenig, H. G. (2002). A Review of Mortality Research on Clergy and Other Religious Professionals. *Journal of Religion and Health, 41*(1), 57–68.

Gardner, J., & Oswald, A. (2004). How Is Mortality Affected by Money, Marriage, and Stress? *Journal of Health Economics, 23*(6), 1181–1207.

Gavrus-Ion, A., Sjøvold, T., Hernández, M., González-José, R., Martínez-Abadías, N., Torné, M. E. E., & Esparza, M. (2021). Religion and Fertility Patterns: Comparison of Life History Traits in Catholics and Protestants, Hallstatt (Austria) 1733–1908. *Journal of Biosocial Science, 53*(2), 305–318.

Giles, T., Hungerman, D. M., & Oostrom, T. (2023). *Opiates of the Masses? Deaths of Despair and the Decline of American Religion* (No. w30840). National Bureau of Economic Research.

Hurwitz, A., Mitchell, O. S., & Sade, O. (2022). Testing Methods to Enhance Longevity Awareness. *Journal of Economic Behavior & Organization, 204*, 466–475.

Idler, E., Blevins, J., Kiser, M., & Hogue, C. (2017). Religion, a Social Determinant of Mortality? A 10-Year Follow-up of the Health and Retirement Study. *PLoS ONE, 12*(12), e0189134.

McCullough, M. E., & Willoughby, B. L. (2009). Religion, Self-Regulation, and Self-Control: Associations, Explanations, and Implications. *Psychological Bulletin, 135*(1), 69.

Milevsky, M. A. (2020). Retirement Income Recipes. In *R: From Ruin Probabilities to Intelligent Drawdown*. Springer-Nature.

Mittal, C., Griskevicius, V., & Haws, K. L. (2020). From Cradle to Grave: How Childhood and Current Environments Impact Consumers' Subjective Life Expectancy and Decision-Making. *Journal of Consumer Research, 47*(3), 350–372.

Pargament, K. I., Koenig, H. G., Tarakeshwar, N., & Hahn, J. (2001). Religious Struggle as a Predictor of Mortality among Medically Ill Elderly Patients: A 2-Year Longitudinal Study. *Archives of Internal Medicine, 161*(15), 1881–1885.

Payne, J. W., Sagara, N., Shu, S. B., Appelt, K. C., & Johnson, E. J. (2013). Life Expectancy as a Constructed Belief: Evidence of a Live-to or Die-by Framing Effect. *Journal of Risk and Uncertainty, 46*, 27–50.

Shor, E., & Roelfs, D. J. (2013). The Longevity Effects of Religious and Nonreligious Participation: A Meta-Analysis and Meta-Regression. *Journal for the Scientific Study of Religion, 52*(1), 120–145.

Wallace, L. E., Anthony, R., End, C. M., & Way, B. M. (2019). Does Religion Stave Off the Grave? Religious Affiliation in One's Obituary and Longevity. *Social Psychological and Personality Science, 10*(5), 662–670.

CHAPTER 3

The Benefits of Pooling

Introduction and Stuff

This chapter gets to the financial and actuarial core of the book; namely, it addresses head-on why the discovery—or perhaps better described as the innovation—of longevity risk pooling (or just "pooling," for short) in the mid-eighteenth century was, and continues to be, so important for financing pensions and annuities in the twenty-first. While the specific details of the Presbyterian annuity fund, or the first successful longevity pool in the mid-eighteenth century, will be discussed at length in Part 2 of this book, here I limit myself to the mechanics.

It's nearly impossible to appreciate the religious roots of longevity pooling—and the beliefs that underpin its success—without understanding its statistical mechanics. Yes, the pedagogy involves a fair bit of probability and finance, more so than what I covered in Chapter 2, but I promise to keep any calculus to a minimum and all equations completely out of sight. What I will do, however, is present a series of hypotheticals or thought experiments to help develop the intuition behind longevity risk pooling. Along the way, I'll introduce some of the common insurance and economic terminology that has taken centre stage in the recent debate around equity and fairness of pension pooling. This background will help readers truly understand some of the challenges, both scientific and ethical, that are associated with "blind" risk pools. I'll explain the "blind" reference later on, towards the end of the chapter, but this

M. A. Milevsky, *The Religious Roots of Longevity Risk Sharing*,
https://doi.org/10.1007/978-3-031-62403-2_3

really is one of the main challenges for the future of longevity risk pooling and pensions, and something I will return to (and conclude with) in Chapter 12 of this book.

With the basic mortality, longevity, and risk foundations (i.e., the probability curve) from Chapter 2 out of the way, I can begin to discuss the historical context of risk pooling—and the first successful attempt to actually pool longevity risk—which begins in Chapter 5 and continues in Chapters 6 and 7. For now, though, I will proceed with a question that might seem rather tangential.

How Much Money Do I Need to Retire?

Imagine that (someone called) Colin, who is somewhere in his early 60s chronologically,[1] is contemplating retiring full-time from the workforce, which of course means that his steady work paycheque will come to an abrupt halt. I would like you to further imagine that Colin isn't entitled to any government or corporate pensions and therefore must create his own personal pension or annuity plan, which today is described as a "decumulation strategy." The noun decumulation has only recently become a proper word in English and means a decrease in amount or value.[2] It is the opposite of accumulation. Colin's plan for the decumulation stage of the lifecycle is that he would like to generate or receive $100,000 every single year for the rest of his life, which is approximately what his disposable income was while working. This is a replacement rate of 100% of his disposable income, which is quite different from 100% of his pre-tax income.[3] Now, whether your typical Colin actually needs the same disposable income as he required while working is completely beside the point and not the objective of this chapter, which is to explain longevity risk pooling. Think of the $100,000 as his financial goal, which really should be expressed in terms of annual income during his decumulation years

[1] That word is used deliberately to distinguish from biological age, which can differ from chronological age and something that will be addressed in the final Chapter 12 in more detail.

[2] See the *Collins Online Dictionary* definition, for example, at https://www.collinsdictionary.com/dictionary/english/decumulation.

[3] Income taxes and deductions are likely to be lower during retirement.

and not wealth at the point of decumulation, which is only a by-product of the income goal.[4]

Moving on, the question I would like readers to ponder—and one that actually befuddles millions of retirees every year—is the following: *How much money should Colin have saved up to the point of retirement (a.k.a. decumulation) in order to generate the $100,000 per year while he is alive to satisfy his financial goal?* What size of "nest egg," to use the colloquial term, does Colin require to achieve this goal? Remember that Colin's remaining lifetime is random, unknown, and stochastic.

Truth be told, the answer to "How much is enough?" can be (made) very complicated, and entire software packages have been designed to solve this problem. Personally, I prefer to take a very simplified approach, at least to begin with. Simplicity isn't free and to pay for the cost of simple I am compelled to make a number of assumptions about Colin, and in particular how he invests his "nest egg" while he waits for decumulation. The money could in theory be invested in any of a thousand different mutual funds, stocks, and investment vehicles. And, who knows how those will perform over the next years and decades. How does one plan or make projections? Let's assume Colin will take absolutely no financial chances with his nest egg. It will be allocated to a safe investment account earning a guaranteed 5% interest rate, each and every year, forever.[5]

Now, you might cringe at the simplicity—or perhaps the appropriate word here is "danger"—of this assumption but recall that this is only a thought experiment, and my objective isn't to give Colin a full decumulation plan. Rather, my limited job is to convince you over these next few pages that Colin might want to (1) longevity pool his financial resources, and, more specifically, that (2) he can do that by acquiring a pension life annuity.

Recall from Chapter 2 and the curve I alluded to, that the length of Colin's life is random. However, in the absence of any further information about Colin's longevity prospects, I'll assume that he's interested in financially planning to live to the age of 85, that is a 20-year horizon. Such

[4] Not to delve into the philosophy of retirement income planning, I would refer the interested reader to Kotlikoff and Burns (2008).

[5] This was the risk-free rate available at the time of writing, that is early in calendar year 2024.

a simple approach—that is, fixing a deterministic retirement horizon—might entirely contradict the main message from Chapter 2, namely, that Colin might live to the age of 100 and perhaps beyond. Yet many financial professionals (and software apps) take an overly simplistic approach to longevity risk by pre-selection or choosing exactly how long they plan to live. Notwithstanding the religious subtext of this book, retirees aren't gods and can't predict their death—even if only for the purpose of a financial planning exercise. Longevity risk is an ever-present factor of retirement life.

Nevertheless, and for the deterministic now, I can declare with full confidence that if Colin would like to withdraw exactly $100,000 per year for only 20 years—and is content with zero if he happens to exceed the two decades—then he must have a nest egg or investment fund of exactly $1,264,000 set aside at the age of 65. His financial goal of $100,000 in annual income can be translated or mapped into a required sum of $1,264,000 at the point of decumulation.

If Colin begins with more than $1,264,000 at age 65, the decumulation withdrawals will last beyond age 85, and if he has less than $1,264,000 at age 65, he will run out of money (a.k.a. get ruined) before age 85. This dollar number—and the confidence behind it—assumes that Colin will continue to make those $100,000 extractions (which is equivalent to 8% of his original nest egg) regardless of anything that happens in his remaining life. A tall order, but well within the bounds of a thought experiment.

That $1,264,000 number obviously wasn't pulled out of thin air, nor was it the result of a terribly complex computer simulation or mathematical calculation. Rather, it was obtained using a very common financial technique known as computing present values. The present value of $100,000 over the next 20 years is precisely $1,264,000, at a 5% interest rate. Yet another way to prove or convince yourself that in fact $1,264,000 is the correct answer is something that I urge my students to do themselves, which is to simulate the path of the account by brute force. How? Start with $1,264,000 and have it earn 5% for the first year and then withdraw $100,000 and then have the remainder earn 5% for the second year, and then withdraw $100,000 for the second year, and continue this growth and withdrawals process until the money runs out. You will see that at age 85, after 20 years, there is nothing left. Ergo, you need $1,264,000 at age 65 to finance 20 years of $100,000 annual withdrawals.[6]

Now, back to the decumulation plan. If Colin is (rightfully) concerned that 20 years is too short a time horizon over which to plan and instead would like to budget or reserve for 30 years (i.e., 10 more years) of withdrawals, then he obviously needs more set aside at age 65. In that case, a more conservative and safe approach would be to increase the appropriate present value to $1,554,000. In that case, the $100,000 divided by the $1,554,000 is an initial withdrawal rate of 6.4%. The methodology for arriving at the $1,554,000 is exactly the same, except that the present-value formula uses 30 years instead of 20. And, if Colin wants to be extremely prudent, and worries about living to age 105, which is 40 years of withdrawals, he needs $1,729,000 at the point he begins decumulation. That is an initial withdrawal rate of 5.8%. On a related note, the reason (some) financial practitioners recommend a lower 4% initial withdrawal rate, when it appears 5.8% gets you 105, is because they are increasing the $100,000 each by inflation. I'm not, yet.

For the record, the oldest person whose age has been verified lived to age 122, but I'll round that down to the biblical 120. So, if Colin wants what I will call biblical security then he must have $1,872,000 set aside to begin with for 55 years of withdrawals. Again, that's an initial withdrawal rate of 5.3%. To be clear, these numbers were computed deterministically with nothing fancier than a present-value (PV) formula, and the only difference between the numerical values come down to the time horizon: 20 years versus 30 versus 40 versus 55.

Table 3.1 displays these present values at the 5% valuation or discount rate, listed in different rows, depending on the terminal age to which Colin plans to live. Namely, if Colin is willing to gamble on a very short 5-year length of remaining life, all he needs upfront is $442,000 at age 65. This is nearly a quarter of what someone would require if he planned to age 100, and entirely depends on how much of a risk Colin is willing to take—or perhaps how long he thinks he might live. Clearly, there is no universal or global sum of money that a 65-year-old should have set aside to finance retirement, or universally accepted initial withdrawal rate, since, at the very least, it depends on longevity assumptions.

[6] For those diligent readers who want to replicate this number for themselves, this assumes continuous compounding, so that the $100,000 is received slowly over time (e.g., weekly), not once per year, and the 5% interest is also earned continuously.

Table 3.1 What do you need at age 65 to withdraw $100,000 per year?

Age to which you plan	*"Nest egg" required ...to that age*	*Probability of living ...to that age (%)*
70	$442,000	94.8
75	$787,000	86.8
80	$1,055,000	75.1
85	$1,264,000	59.2
90	$1,427,000	39.9
95	$1,554,000	20.9
100	$1,652,000	7.2
105	$1,729,000	1.2
120	$1,872,000	$\ll 0.05$

All Money Earns 5% per year while invested and unused.
Survival Probabilities based on Gompertz Approximation.
See Chapter 7 for more information about mortality models

Thus, I have explained the second column in Table 3.1. Moving across, the third column in Table 3.1 will show and warn Colin that planning withdrawals only to age 85 is a rather risky business. There is a near 60% probability that he will still be alive at age 85, and if he only budgeted for 20 years of retirement, his nest egg would be empty. Another way of expressing this is that there is a 40% probability (well, 0.408 to be precise) that he will live to the age at which the money is depleted, that is age 85. Moving down the rows, there is a 20.9% probability that Colin will live to age 95. Once again, if he only plans to live to age 95—and begins with a $1,554,000 nest egg—there is a 20.9% chance that he will run out of money. Why? Because the money will run out at age 95 and there is a 20.9% chance that he will be alive, ergo there is a 20.9% chance that he will run out of money.

At the risk of overexplaining this rather important point, here is yet another way to interpret the numbers in Table 3.1, which will become the foundation of any longevity risk pooling example. *If* Colin retires at the age of 65 with a $1,652,000 nest egg, and *if* he invests that money in a fund that earns 5% interest per year, and *if* he insists on spending $100,000 from that fund every year—yes, that is a lot of *ifs*—there is a 7.2% chance he will run out of money before he runs out of life. This number is obviously less than a 1-in-10 chance, but more frequent than a 1-in-100 chance. A financial practitioner might subtract 100% from

7.2%, and declare a 92.8% success (versus failure) rate with a $100,000 annual withdrawal under a guaranteed 5% investment return.

To sum up the main "decumulation philosophy" underlying Table 3.1, whether Colin thinks that $1,652,000 is enough money at age 65 really comes down to his attitude to—and fear of—longevity risk. The longer he wants to plan for—despite the lower odds—the more he needs up front. Now, to further compound the risks by generalizing the example, what if Colin can't earn 5% on his invested money? What if the safest interest rate is only 4% or 3%? In that case, $1.6 million is even less likely to be enough. The point here is that there is no single, absolute, and known "number" that solves the retirement (a.k.a. decumulation) dilemma: *How much do I need?* It comes down to probabilities and attitudes.

What Exactly Do These Living Probabilities Mean?

Before I move on to what "longevity risk pooling" can do to help Colin, I will offer a bit of context to explain where these living probabilities (third column in the table) come from as well as how they should—and shouldn't—be interpreted. First, to be very clear, human life is not a perfectly designed and symmetric casino roulette wheel, where the probability of getting 10 reds in a row can be computed precisely as $(18/38)^{10} = 0.0128\%$, when there are 18 reds, 18 blacks, and 2 green slots on the wheel. Yes, spend a lifetime in a casino spinning roulette wheels 10 rounds at a time and you can be assured that by the end of your life, 0.0128% of those trials will have all been red. That is the only context in which the word probability can be used with absolute certainty and with no disclaimers or adjectives, such as estimated probability, approximate probability, or likely probability. Remember this when someone predicts the next winner of the FIFA World Cup with a probability estimate such as "Argentina will win with a 60% probability." The word probability is being abused, and I'll return to the historical meaning and eighteenth-century context of probability in Chapter 9.

Similarly, real biological life isn't nearly as neat and subject to modelling as mechanical systems. Therefore, in a run-up to the discussion of probabilities in the eighteenth-century Presbyterian fund, this is a good place to remind readers that any probability of survival—such as the ones in the third column of Table 3.1—is only an estimate based (partially) on historical data observed over long periods of time. I might have data that indicate that 59.2% of humans who ever made it to the age of 65 (in the past) have then gone on to survive or live to the age of 85. One can then extrapolate from large groups (in the past) to single individuals (in the present) that someone who is 65 today has a 59.2% probability of reaching age 85, assuming no change in the mechanisms driving mortality.

Notice the subtle difference between a demographic historians declaring that 59.2% of a very large group of people *in the past* had gone on to survive 20 more years versus a forward-looking actuary or statistician predicting or prophesizing that you or I have a 59.2% probability of living to age 85. Yes, these sorts of numbers were implicit (and often confused) in the discussion of longevity risk in Chapter 2, and in particular Fig. 2.1. Notice the difference between a historical frequency versus a forward-looking forecast. In fact, students who are new to actuarial science are often puzzled by these types of very precise statements. They often ask: *What does it even mean to be 59.2% alive or 40.8% dead?* Answer: Large group of people, frequency of survivors.

To be clear, I don't want to cast doubt on these probability numbers but rather to remind readers that (1) the proper way to think about these probabilities is as a frequency or a fraction applied to a large homogenous group at the same age and (2) these forward-looking projections contain quite a bit of uncertainty themselves. A statistician might accurately pronounce that the probability of a 65-year-old surviving to 85 is between 55 and 65%, or that there is a ±5% confidence range around the 59.2% point estimate, and so on.

The point here isn't to get bogged down in the statistical minutia of mortality forecasting but rather to emphasize the hypothetical nature associated with these examples and numbers. When I write (per the last row in Table 3.1) that $1,729,000 might not be a large enough nest egg at retirement to generate $100,000 per year because there is a 1.2% probability Colin will still be alive at age 105 when the money runs out, keep these caveats in mind. In fact, the farther out one goes in time—that is, the farther in the future I try to predict whether Colin will be alive—the greater the uncertainty with missed things like medical breakthroughs or global pandemics or anything in between. This is why the very last row in Table 3.1, with age 120 listed as the terminal horizon, has an inequality (less than 0.05%) instead of a precise number. It's a reminder. I have no idea—and I doubt anyone does—what that number truly is. Historically, very few have made it to that age, which makes extrapolating nearly impossible.

Couple Plans for Retirement and Pools

Let me recap the financial and actuarial risks that Colin faces. If he wants to withdraw $100,000 per year for the rest of his life and also wants 90% "probability confidence" that he will have enough money for that objective, then he should begin his decumulation plan for retirement with $1,629,000 in savings. A 90% success rate is a 10% failure rate, which might be too high for Colin. So, the same table would also inform him that if he wants a much higher 99% probability confidence—that is, being successful in 99 out of 100 scenarios—then he must begin at the age of 65 with $1,735,000 in financial resources for retirement. In fact, those two numbers are in some sense repeated in the first row of the following Table 3.2, which reflects the probability of living to the age at which the money runs out. Now, it's finally time to explain the single most important actuarial idea underlying this chapter and, in some sense, this book: It's called pooling.

Now imagine that Colin joins forces with (someone called) Anne—in one large pot—to collectively fund and finance their retirement decumulation. They each want to withdraw $100,000 from their nest egg, which

Table 3.2 The financial benefits of pooling resources

Size (N) of longevity risk pool at age 65	*Reserve fund "nest egg" required at age 65, to sustain $100,000 withdrawals every year: $ amount (% cost reduction)*	
	Confidence of 90%	*Confidence of 99%*
1 (no pooling)	$1,629,000 (N.A.)	$1,735,000 (N.A.)
2	$1,549,000 (−4.9%)	$1,662,000 (−4.2%)
3	$1,504,000 (−7.7%)	$1,617,000 (−6.8%)
10	$1,393,000 (−14.5%)	$1,486,000 (−14.4%)
100	$1,292,000 (−20.7%)	$1,329,000 (−23.4%)
500	$1,266,000 (−22.3%)	$1,283,000 (−26.1%)
5000	$1,251,000 (−23.2%)	$1,257,000 (−27.6%)

Note As in Table 3.1, the money earns 5% every year with certainty
Hypothetical cost of life annuity is $1,243,000

is a total of $200,000. How much do they need together in their larger fund to generate the $100,000 per year for each one? This could also be expressed as a question: *Can they increase their initial withdrawal rate if they do it together versus alone*? Note, this isn't about living together and reducing their living expenses due to economies of scale, or that buying in bulk wholesale is cheaper than buying retail. No, Colin and Anne can be living on opposite sides of the world. They each want to extract $100,000 from their shared communal fund.

This question—*How much do they need upfront?*—is at the core of longevity risk pooling and your initial reaction might be to (lazily) double the numbers. If they each want 90% confidence, then parallel with the $1,629,000 per person, you would think they should set aside $3,258,000 as a couple. But alas, that answer is quite wrong and the essence of pooling. Insisting on a fund-for-two with $3,258,000 jointly (or $1,629,000 individually) overstates the money or the reserves the couple requires to finance their joint retirement under the same 90% success probability. Why? Here is the key ingredient: If one dies (early), the other inherits the remaining money.

So, why can't you just double the reserve when you double the people? Well, remember that as a married couple, Colin and Anne likely have an implicit agreement to support each other with their funds. This means that if and when the first of them happens to die, whether it is Colin or Anne, the survivor will inherit the assets of the deceased. That's exactly what it means to pool resources, although this is arguably a very tiny pool. This windfall gain—for lack of a better term—will further help sustain the survivor's retirement withdrawals. Another name, term, or phrase for the extra money that goes to the second person when the first one dies is what actuaries call "mortality credits." Colin or Anne will earn these mortality (some call it longevity) credits as soon as the other one passes away.

It is money, but it can be viewed or thought of as a type of subsidy, interest rate, or just bonus credit. Either way, knowing that the survivor will inherit the funds of the deceased—whichever one it might be or whatever it is called—means they both can set aside less money at the very beginning, at age 65. Alternatively, they can increase their initial withdrawal rate. They don't have to wait for the death to happen, because they know that eventually one of them will die (first) and they can plan to have just a little bit less now for the two of them. Yes, this might be a very morbid and harsh way of thinking about Colin and Anne living out their remaining years in retirement, but the point is to explain how their joint longevity risk is reduced and why *all together* they need less to finance their retirement.

The second row in Table 3.2 tells us exactly how much Colin and Anne must each set aside today if they plan to pool their resources for retirement together. More specifically, if they each would like the same 90% probability confidence (or 10% failure rate) of achieving their retirement withdrawal goals, they each must contribute $1,549,000. Again, it is lower than the *go-it-alone* number of $1,629,000, that is, if they managed their financial affairs alone and unpooled.

That's a savings of $80,000 or almost 5% (nothing to do with interest rates or survival probabilities, although they both affect it) in resources required to finance their joint retirement. To be very clear yet again, these savings have absolutely nothing to do with the economies of scale—cooking for two people is less than double the cost of cooking for one, and so on—that come from sharing expenses and living together. Yes, it's also quite likely they each need less than $100,000 per year when they pool their expenses, and I don't even want to get into a discussion of the tax benefits (and savings) of being married. That's not what Table 3.2 is about, and that's not the benefit of longevity *risk* pooling. It just adds to the pooling benefits. What I have just described is a mathematical benefit based purely on probability theory, not personal finance or home economics or tax policy. If one member of a couple (or a pool of two) is willing to forfeit any remaining assets to the other when they die, they both can maintain a higher standard of living when they are alive. Stated differently, they need less initially, at retirement, to achieve the same financial goals with the same level of statistical confidence.

Of course, if Colin and Anne want 99% (versus a mere 90%) statistical confidence to achieve their individual financial goal of withdrawing $100,000 per year, their reserve fund at age 65 must be higher, but the benefit of pooling resources remains regardless of—what an economist might call—their level of longevity risk aversion. As you can see from the third column of Table 3.2, if they decide to go it alone and desire 99% confidence, they must begin their retirement journey with $1,735,000 versus the $1,629,000 under 90% confidence. But, once again, if they pool their financial resources, that amount drops by 4.2% to $1,662,000. The benefits to longevity risk pooling are similar regardless of the statistical confidence one uses or, once again as an economist would say, regardless of their level of risk aversion and tolerance. And, if you aren't impressed by the nearly $80,000 reduction in the cost or price tag for retirement, you should note that these assume a guaranteed 5% interest per year. If you don't think that can be achieved, that is if you worry that 4% or perhaps 3% is more realistic, you should note that the gains from longevity risk pooling are even greater because the cost of retirement is higher.[7]

Three Is a Crowded Pool, But Is Even Better

With that pool-of-two logic firmly behind us, let's expand our couple into a larger group and get closer to a true longevity risk pool. Assume now that Colin and Anne find a third person to join their union—and I mean financial union—so that a total of three people are pooling their financial resources. Again, the financial nature of the agreement is the same. They combine their retirement reserve funds into one large pool and start to draw $100,000 each or $300,000 in total every year from the larger fund. If and when the first one dies, whether it's Colin, Anne, or the new member of the threesome, the remaining two will continue to withdraw $200,000 per year, and then after the second one dies, the last survivor will withdraw $100,000 until he or she dies. I have no idea when these deaths will take place—remember the randomness of life per Fig. 2.1—but according to the numbers computed and displayed in Table 3.2, each member of the threesome will have to enter the financial relationship at age 65 with $1,504,000, to maintain the same 90% probability confidence. And, if they each want 99% statistical confidence, their financial contribution to the longevity risk pool will have to be $1,617,000. The most important thing to notice is that both of these numbers (that is money required) are approximately 7% lower than had each of them (or Colin) gone it alone. That's the benefit of longevity risk pooling, measured by shrinking sums of required reserves with the growing size of the pool. This "discovery"—or at least let's call it a pooling insight—was

[7] There are many (recent) attempts to explain the benefits of longevity risk pooling, a.k.a. the magnitude of "mortality credits" within life annuities, and once recent practitioner-oriented article is by Michel Finke, published in October 2023 via the *Retirement Income Institute*, and available here. Another older article (2017) by Joe Tomlinson is available from VettaFi Retirement Perspectives, and yet another explanation (2019) is provided by Daniel Bauer, from the Society of Actuaries. So, different authors and academics have their preferred way of explaining the benefits, but they are all saying the same thing. See also the next section on utility benefits, which is yet another (more abstract) approach.

first implemented successfully in the eighteenth century in Edinburgh, Scotland, by the Presbyterian Church.

Now, the longevity pools I'll be discussing in this book—and the ones that originated in the eighteenth century—are far greater than two or three people and involve tens if not hundreds of participants who pool mortality and longevity risk. The key here is to notice how, as the number of participants in the pool increases in Table 3.2, the magnitude of the required fund at time zero—which again is just the cost of retirement for each member—plummets, while still maintaining the same standard of living while alive. In fact, once the pool size reaches 5000 people, the cost reductions amount to between 23 and 28% of the go-it-alone cost, depending on the level of probability confidence they would like to achieve. In fact, when interest rates are lower (than 5%) the pooling benefit is even greater, relatively speaking. Note the diminishing marginal utility—to use a term from microeconomic theory—from larger and larger pools. The benefit of pooling doesn't grow to infinity. Notice how in Table 3.2 going from a pool of 100 people to a pool of 5000 people doesn't reduce the required initial retirement (a.k.a. decumulation) fund by very much. Yes, there might be other benefits to much larger pools—such as reduced operating expenses and lower transaction costs—but those matters are separate from the longevity risk element.[8]

In fact, pooling longevity risk not only reduces the discounted cost of financing a stream of withdrawals during retirement, it also reduces the dispersion or uncertainty associated with the cost as well. To understand this rather subtle insight, think of the range of outcomes when you don't pool longevity risk versus the situation when you choose (or are required) to do so. If you wanted a 90% probability confidence interval for how much money you need to finance your \$100,000 retirement at age 65 if you go it alone, the answer would be somewhere in the range of \$432,000 (die early) and \$1,673,000 (die late). This echoes the numbers displayed in Table 3.1, where success and failure solely depend on how long you alone live. But, for the same 90% confidence interval when you pool with 50 people, the range narrows to between \$1,154,000 and \$1,330,000. The random number of outcomes that can occur—or more precisely, the initial investment fund or life annuity reserves that are

[8] For this empirical debate, namely whether larger pension (a.k.a. life annuity) funds achieve better investment performance over time, see the articles by Bikker and De Drue (2009), Bikker (2017) and Dyck and Pomorski (2011).

required to pay everyone their $100,000 per year—shrinks (dramatically) with the growing size of the longevity pool.[9]

If you haven't guessed this yet, the above is precisely how (modern) insurance companies are able to promise or guarantee life annuity payments to their customers. They pool them together, estimate how long each one of those pool members might live, and then set a payout rate that keeps the overall solvency level (ruin probability) of the entire company at some very high (or very low) number. So, if you walk into an insurance company at the retirement age of 65, with (approximately) $1,250,000 in cash when interest rates in the market are 5%, the company will guarantee you $100,000 per year for life, by pooling your money with everyone else. Had you tried to manage the money alone (without pooling) and tried to extract the $100,000, the money would last somewhere into your early 80s, and certainly not beyond age 85.

Measuring Economic Utility Benefits

What I have just described is called a "probabilistic" approach to explain or show the benefits of longevity risk pooling, otherwise stated as irreversibly annuitizing your liquid wealth. It's irreversible since we obviously don't want participants to leave the pool and take their funds, perhaps just before they die.

Over the years, though, pension and insurance economists have devised other approaches to measure and display the benefits of longevity risk pooling, using more sophisticated models of consumer preferences. These more economic-based (versus probability-based) approaches don't change or modify the basic insight that longevity risk pooling is a good idea and enables retirees to consume or spend more than they would have otherwise. Nevertheless, these economists have pointed out that the magnitude of the benefits can be subjective and depend on risk preferences. Allow me to explain the subjective and risk preference aspects.

[9] For readers who want to replicate these numbers, the numerical results were obtained as follows. First, one million random lifetimes were simulated for a specific pool size (1, 2, 3, 10, etc.) vector, using a standard Gompertz specification with a modal value of 90 years and a dispersion coefficient of 10 years. Then, the present value of $100,000 was computed under a 5% interest rate for each life in the vector. The numbers displayed are the quantiles (at 10% or 1%) of the stochastic present value, divided by the size of the pool.

For example, if you would rather your financial assets go to your children or loved ones after you die—that is, if you have strong bequest motives—and not the anonymous members of the pool, you might be willing to forgo the benefits of longevity risk pooling. Likewise, economists point out that the more (longevity) risk-averse retirees are—that is, the greater their fear or worry about having to reduce their standard of living if they live longer than expected—the more they value risk pooling, and the less they will spend for fear of running out of money.[10]

In one of the earliest economic attempts to compute the utility benefits of longevity risk pooling, economists Larry Kotlikoff and Avia Spivak wrote the following in an article published in the *Journal of Political Economy* in 1981:

> In the case of marriage both individuals commonly agree to pool their resources while both marriage partners are alive and to name each other as the major, if not the sole, beneficiary in their wills. For each partner the risk of living too long is somewhat hedged by the other partner's potential death; if one partner lives to be very old, there is a high probability that his (or her) spouse has already died leaving him a bequest to help finance his consumption. While each spouse gains simply from the exchange of wills, the two can further increase their expected utilities by agreeing on a joint consumption path that takes into account each spouse's expected bequest to the other.[11]

The above quote should resonate with the numbers in Table 3.2, and in particular, the second row in which the single person can reduce the financial cost of his or her retirement by close to 5% by pooling resources with another person. Once again, this is just another way of expressing the same idea.

[10] For more on the idea of longevity risk aversion (versus risk tolerance) see the article cited as Milevsky and Huang (2011). On the topic of longevity risk aversion and how this and other preference parameters might impact optimal (or observed) decumulation rates, a.k.a. portfolio withdrawal rates (PWR), see the work cited as Asebedo and Browning (2020), Andréasson et al. (2017), Dang et al. (2017), Konicz and Mulvey (2015), Pfau (2018), Scott et al. (2022), as well as Sun and Webb (2013.).

[11] Page 380 of Kotlikoff and Spivak (1981), which is one of the most widely cited papers in the field that attempt to measure the benefits of longevity risk pooling. Another important article that builds on the Kotlikoff and Spivak (1981) methodology, is Brown (2001).

Now, to be clear, Kotlikoff and Spivak (1981) estimated larger and greater benefits to couple pooling (i.e., marriage of two), with estimates as high as 25% of wealth, because they use a different economic methodology, as well as lower life expectancy values and inflation-adjusted interest rates. In fact, if I reproduce Table 3.2 uses a life expectancy that is reduced by 10 years, which leads to a lower survival probability, and an interest rate that is reduced by 4% inflation. Then, the couple's benefit jumps to 12% savings in wealth (versus 4.9%) and pooling with 99 others reduces retirement costs by 36% (versus 20.7%). And, while I don't want to get bogged down in econometric assumptions, as I have hinted earlier, reducing the interest rate "assumption" increases the magnitude and benefit of longevity risk pooling because (1) consumption itself is more expensive, and therefore (2) the mortality credits are relatively more valuable.[12]

Abstracting from the details, all economists agree that (some) longevity risk pooling is beneficial. The only debate revolves around the best model to illustrate benefits as well as the magnitude of the gains.

In fact, Kotlikoff and Spivak (1981) have been cited many hundreds of times and their methodology has become the gold standard in the pension economics literature. And, although they use a rather complex "dynamic programing" algorithm to compute "discounted lifetime utility," their qualitative results are entirely consistent with the message emanating from Table 3.2. Life becomes more economical the more you pool. The authors estimate how much (less) money someone would need—to generate an identical level of "economic utility"—if they were willing to marry (share) with more than one person. Yes, you read that correctly. They estimate pooling benefits of a three-person polygamy (p. 383) and a four-person mixed marriage consisting of two 55-year-olds together with two 30-year-olds (p. 386). In this case, if they all contributed the same resources, the 50-year-olds would be implicitly subsidizing the two 30-year-olds, which is a complex (ethical) matter that I'll get to in a moment. For now, I'll note that the longevity risk pooling benefits to a three-person or four-person "partnership" are quite substantial, and for safety's sake I'll stop here.[13]

[12] On this note, the cost of retirement consumption when real interest rates are lower, see also Blanchett et al. (2017).

[13] Note that Kotlikoff and Spivak's article (1981) was published over 40 years ago, perhaps when academic editors had a better sense of humour and/or were less sensitive to cultural matters. See Milevsky and Huang (2019) for extensions of the Kotlikoff and

Mixed Pools and Mortality Heterogeneity

Notwithstanding the (partially humorous) comments by the two economists, their ideas about mixed marriages—or longevity risk pools with members of different ages—do raise a sensitive point. It is a matter that I will return to in the final chapter. In brief, though, to understand the problem with mixed pools, let's return to Colin and Anne, where I began the discussion of pooling. While most of this chapter revolves around hypothetical and idealized examples—including the polygamy quip—it's very likely that in real life Colin (male) is older than Anne (female). More importantly, what this means from an actuarial point of view (and think back to Fig. 2.1 in Chapter 2), Anne's life is likely to be longer than Colin's, perhaps even by decades. His (or any male's) longevity prospects and probabilities are lower, even if they are at the same chronological age today. While the numbers underlying the hypothetical calculations in Table 3.2 assume longevity homogeneity or a complete symmetry of risks, reality is unlikely to be as neat. If Colin is not expected to live longer—and again I urge you to look back at Fig. 2.1 once again, which displays the probability distributions of remaining lifetimes—then Anne appears to be benefiting much more from pooling as compared to Colin. After all, she is more likely to inherit his leftovers versus the other way around. This is more than a theoretical wrinkle in the Colin and Anne pooling argument and actually became a rather contentious issue near the dawn of longevity risk pooling in the eighteenth century when it appeared that older (pool) participants were taking advantage of their younger members. This will be explained in due time in the second and third parts of the book. And, as I delve into the historical concerns about pooling fairness, I'll show how that was partially addressed by appealing to religious arguments, but at this point, I'll note that this is a growing problem in the twenty-first century as well. It's quite clear today that some members and groups within society live (much) longer than others, and this isn't entirely random and unpredictable, and yet they are often pooled together in the same pension plan with much healthier people who are likely to live much longer.

Spivak (1981) model, as well as the classic article Yaari (1965). For additional (and recent) research on the "utility value" of annuitization, see the following articles: Alvarez et al. (2021), Abou Daya and Bernard, C. (2022), Balter et al. (2021), Bernard et al. (2021), Chen et al. (2021), Kirusheva et al. (2022), as well as Steffensen and Søe (2023), in their article suitably entitled: *What is the value of the annuity market?*

For now, the key takeaway from this chapter is how longevity risk pooling works in practice; how it relates to the design of life annuities which promise a pension for life; the different ways to measure those benefits; and how it makes retirement (and decumulation) financing cheaper. We are now ready to head back in time and visit the historical archives, to examine the first successful attempt to pool longevity risk and offer annuities for life. Surprisingly, the genesis can be found in both a church and a university.

Sources and Additional References

Abou Daya, M. H., & Bernard, C. (2022). What matters in the annuitization decision? *Swiss Journal of Economics and Statistics, 158*(1), 14.

Alvarez, J. A., Kallestrup-Lamb, M., & Kjærgaard, S. (2021). Linking retirement age to life expectancy does not lessen the demographic implications of unequal lifespans. *Insurance: Mathematics and Economics, 99*, 363–375.

Andréasson, J. G., Shevchenko, P. V., & Novikov, A. (2017). Optimal consumption, investment and housing with means-tested public pension in retirement. *Insurance: Mathematics and Economics, 75*, 32–47.

Asebedo, S. D., & Browning, C. M. (2020). The psychology of portfolio withdrawal rates. *Psychology and Aging, 35*(1), 78.

Balter, A. G., Kallestrup-Lamb, M., & Rangvid, J. (2021). Macro longevity risk and the choice between annuity products: Evidence from Denmark. *Insurance: Mathematics and Economics, 99*, 355–362.

Bernard, C., Aquino, L. D. G., & Levante, L. (2021). Optimal annuity demand for general expected utility agents. *Insurance: Mathematics and Economics, 101*, 70–79.

Bikker, J. A. (2017). Is there an optimal pension fund size? A scale-economy analysis of administrative costs. *Journal of Risk and Insurance, 84*(2), 739–769.

Bikker, J. A., & De Dreu, J. (2009). Operating costs of pension funds: The impact of scale, governance, and plan design. *Journal of Pension Economics & Finance, 8*(1), 63–89.

Blanchett, D., Finke, M. S., & Pfau, W. D. (2017). Planning for a more expensive retirement. *Journal of Financial Planning, 30*(3), 42–51.

Brown, J. R. (2001). Private pensions, mortality risk, and the decision to annuitize. *Journal of Public Economics, 82*(1), 29–62.

Chen, A., Hieber, P., & Rach, M. (2021). Optimal retirement products under subjective mortality beliefs. *Insurance: Mathematics and Economics, 101*, 55–69.

Dang, D. M., Forsyth, P. A., & Vetzal, K. R. (2017). The 4% strategy revisited: A pre-commitment mean-variance optimal approach to wealth management. *Quantitative Finance, 17*(3), 335–351.
Dyck, I. J., & Pomorski, L. (2011). Is bigger better? Size and performance in pension plan management. Size and Performance in Pension Plan Management *Rotman School of Management Working Paper* (1690724).
Kirusheva, S., Huang, H., & Salisbury, T. S. (2022). Retirement spending problem under Habit Formation Model. *arXiv preprint* arXiv:2210.06255.
Konicz, A. K., & Mulvey, J. M. (2015). Optimal savings management for individuals with defined contribution pension plans. *European Journal of Operational Research, 243*(1), 233–247.
Kotlikoff, L. J., & Burns, S. (2008). *Spend'Til the end: The revolutionary guide to raising your living standard—Today and when you retire*. Simon and Schuster.
Kotlikoff, L. J., & Spivak, A. (1981). The family as an incomplete annuities market. *Journal of Political Economy, 89*(2), 372–391.
Milevsky, M. A., & Huang, H. (2011). Spending retirement on planet Vulcan: The impact of longevity risk aversion on optimal withdrawal rates. *Financial Analysts Journal, 67*(2), 45–58.
Milevsky, M. A., & Huang, H. (2019). The utility value of longevity risk pooling: Analytic insights. *North American Actuarial Journal, 22*(4), 574–590.
Pfau, W. D. (2018). An overview of retirement income planning. *Journal of Financial Counseling and Planning, 29*(1), 114–120.
Scott, J. S., Shoven, J. B., Slavov, S. N., & Watson, J. G. (2022). The life-cycle model implies that most young people should not save for retirement. *The Journal of Retirement*. https://doi.org/10.3905/jor.2022.1.119
Steffensen, M., & Søe, J. B. (2023). What is the value of the annuity market? *Decisions in Economics and Finance*, 1–26.
Sun, W., & Webb, A. (2013). Should households base asset decumulation strategies on required minimum distribution tables? *The Geneva Papers on Risk and Insurance-Issues and Practice, 38*, 729–752.
Yaari, M. E. (1965). Uncertain lifetime, life insurance, and the theory of the consumer. *The Review of EcOnomic Studies, 32*, 137–150.

PART II

Church & University

CHAPTER 4

An Enlightened Financial Innovation

Summary of Part One

If I can summarize the three main takeaways from the prior three chapters in Part 1 of this book, it would be as follows: First, it seems that individuals who are affiliated with some sort of religious faith group or communal organization live longer than others who don't "believe." However, all groups in society face longevity risk, which captures the uncertainty around how long we live. Second, the financial implications of longevity risk can be mitigated—that is, the cost of retirement can be reduced—by pooling and sharing resources with members of a group; any group. Third, the largest implementation of longevity risk pooling takes place within national pension schemes, and oddly enough there seems to be a correlation between the quality and benefit of a nation (or groups) pension scheme and whether they are members of a Protestant Christian denomination. So, with those three insights established, I am now ready to go back in time to the first-ever attempt to successfully pool longevity risk, which was implemented by a group of Protestant Christian ministers in the eighteenth century, during a period known as the Scottish Enlightenment.

M. A. Milevsky, *The Religious Roots of Longevity Risk Sharing*,
https://doi.org/10.1007/978-3-031-62403-2_4

Introduction to History

Oxford University Professor Hugh Trevor-Roper (b. 1914, d. 2003), one of the most famous and well-regarded historians from the twentieth century, perceived the Scottish Enlightenment to be an event limited to a handful of social and political commentators within Scotland during the mid-eighteenth century. To him, it was a limited affair. In contrast to this narrow view, Professor Alexander Broadie in the preface to his book *The Scottish Enlightenment*, asserts that his "sympathies are with the inclusivist" who also embraces mathematics and natural sciences as being part of the Scottish Enlightenment phenomenon. Brodie highlights the contributions to Newtonian mechanics of Colin Maclaurin (b. 1698, d. 1746), as well the development of non-Euclidian geometry by Thomas Reid (b. 1710, d. 1796), as two examples of the scientific aspects of the Enlightenment. Not to distract from the main point here, both Maclaurin and Reid will make an appearance in the documents of the 1744 fund. For now, I will just note that they were key contributors to the Scottish Enlightenment. Yet another inclusivist—namely, that the Enlightenment is broader than the liberal arts—is Professor Roger Emerson. He wrote that he "shares the view that any adequate understanding of the Scottish Enlightenment must give a larger place to...science." He laments that "science...counted for more...than is usually apparent in the accounts given of it."[1]

This chapter is a historiographic review—focusing on what others have written about the period in question—which expands on this "inclusivist" view. In particular, I argue that the period known as the Scottish Enlightenment should also be identified as the intellectual origins of a key milestone in actuarial science, namely, the first successful attempt to pool longevity risk and properly fund life annuities. Of course, this innovation in thinking wasn't implemented within a "liberal and enlightened" or University academic environment, but rather within the formal Church of Scotland. But, those readers who are familiar with this historical period will know those two groups—members of the church and the university—worked together and respected each other (for the most part), and certainly cross-fertilized intellectual trends in the mid-eighteenth

[1] Trevor-Roper as quoted in Sher (1985, p. xv), Brodie (2007, preface), and Emerson (1988).

century, in ways that would seem very counter intuitive in the twenty-first century. In fact, the so-called conflict between religion (church) and science (university) in itself is something of a myth, and something I'll return to in Chapter 9, as well.[2]

As noted in Chapter 1, in the year 1744 the world's first scheme for paying life annuities to surviving widows and orphans—the innovation at the core of this book—was launched by the Church of Scotland. Although annuities had been available for millennia, the scheme was unique because its managers both computed and established actuarial reserves based on the nascent "doctrines of probability." This financial innovation has been noted by (mass) popular historians such Yuval Halevi and Niall Ferguson in their respective (bestselling) books, albeit with some inaccuracies and omissions.[3]

I'll remind readers that practicing actuaries—especially those of Scottish origin—are conscious of this pioneering development, but their writings have treated Edinburgh circa 1744 as just another date in the eighteenth century, somewhere between Edmond Halley's 1693 mortality tables and the 1772 work on reversionary annuities by Richard Price; both Halley and Price were living in London. The actuarial recognition for the plan has not been embedded within the broader trend known as the Scottish Enlightenment.[4]

In contrast with the actuaries, scholars, and historians of the Scottish Enlightenment appear to have neglected this scheme, for the most part. And, if the scheme is occasionally noted, it's a footnote to Colin Maclaurin's scientific contributions. In fact, his *Oxford Dictionary National Biography* entry, although noting his mathematical work on what are known as (now obsolete) fluxions, the biographer makes no mention of

[2] See the work cited as Sher (2015), which is the revised edition of his original 1985 book, and the inspiration for the title of Part 2 of this book.

[3] Harari (2014), Chapter 14, and Ferguson (2008), Chapter 4. The inaccuracies and omission will be footnoted when the scheme is explained. For one, both authors confuse the fund with the (unrelated) insurance company *Scottish Widows,* which was only founded much later, in the year 1815.

[4] See, for example, the book by Lewin (2003), the book reviews by Sibbett (2014), the keynote address by Deucher (1896), as well as Dow (1975), all practicing actuaries. For more on the history of actuarial science and practice, and in particular with reference to Scotland, see the early article by Patricia Wilkie, the review article by Devadeep Gupta, as well as (anything written by) Chris Lewin and David Bellhouse.

his contribution to a scheme that lasted 250 years after his death, of which he was a key contributing member.[5]

In this chapter, and more broadly during the next few chapters, I will argue that both the mathematical foundation and practical management of the longevity risk pooling scheme really should be placed within the wider context of the Scottish Enlightenment and deserves to be part of that wider cultural phenomenon.[6] This scheme—formally the 1744 Presbyterian Annuity Fund[7]—had roots in both the church and the university, was managed by individuals active in the Scottish Enlightenment, was created within the spirit of economic improvement, and employed probabilistic ideas percolating in the writings of David Hume, Adam Smith, and other literati. It's simply too close in time and place to be a coincidence or a sideshow to the Scottish Enlightenment.

Caring for Widows Over the Ages

Of course, the proper care for widows and orphans has a long-standing tradition of importance with Christianity; Ambrose of Milan (b. 339, d. 397) noted their exalted status in *Concerning Widows*. He wrote: "How great is the gift of divine blessing in widows," which has echoed their status and literary treatment, even if they were financially destitute and the least fortunate members of society. Their suffering, so to speak, was noble.

But, after the Protestant Reformation, which was initiated by Martin Luther and like him allowed clergy to marry, widows posed a unique challenge to the new church. In his *First Book of Discipline* (1560), John Knox—who effectively founded the Church of Scotland—noted that provision must be made for ministers of the Church of Scotland "not only for their own sustenance during their lives, but also for their wives and children after them."[8]

[5] See Bruneau (2020), Grabiner (2002), who do mention his work on the 1774 scheme, as well as Sageng (2004) who does not mention this in his 7-page *ODNB* entry.

[6] To be clear, Niall Ferguson labels the creation of the scheme a milestone in financial history but neglects to place this event within the wider Scottish Enlightenment.

[7] It appears under various names, some of which (like Harari and Ferguson) are confused with the insurance company Scottish Widows.

[8] See the Free Presbyterian Church of Scotland, https://www.fpchurch.org.uk/about-us/important-documents/the-first-book-of-discipline-1560/.

In fact, John Knox himself, in his last moments of life, expressed concern for the future well-being of his own wife, Margaret, and their three daughters, and one of the first acts of the General Assembly of the Church of Scotland upon his death—think of the General Assembly as the main group running the church, like a modern corporation's board of directors—was to assign his widow extra payments.[9] By 25 March 1744, all ministers of the Church of Scotland were required to make annual contributions during their lifetimes, which, upon their demise, entitled their widows to a life annuity or a lump sum benefit to any surviving orphans.

I'm careful not to get ahead of myself and will review the precise details and minutia of the scheme in the next chapter. Again, the 1744 annuity scheme is associated with the names of Reverend Alexander Webster (b. 1707, d. 1784) and Reverend Robert Wallace (b. 1697, d. 1771), with technical assistance provided by Edinburgh mathematician Colin Maclaurin (b. 1698, d. 1746), and (eventually) Reverend Richard Price (b. 1723, d. 1791). These are names you will hear repeatedly when I describe the scheme, and for good reason. But for now, let's examine the scheme's admission criteria.[10]

The Westminster Confession of Faith

In addition to the scheme being open to all ministers in the Church of Scotland, it was also opened to all professors and regents at the major Scottish universities, but provided they accepted the Westminster Confession of Faith (WCF.) To get into the annuity scheme, a.k.a. the longevity risk pool, they had to accept the WCF. What is that? The WCF is a cornerstone of Reformed Christian theology, formulated by the Westminster Assembly, which met between 1643 and 1649. It was created as part of the effort to reform the Church of England and establish a more Presbyterian Church governance. The WCF embodies a comprehensive range of theological doctrines from what is known as a Calvinist perspective, including the sovereignty of God, the authority of the Scriptures, and the nature of the covenant of grace. Central to the WCF is the emphasis on

[9] Dawson (2015, p. 311), and De Romestin (2016), *Concerning Widows*. See also Thane (2020), Kahl (2005).

[10] The most comprehensive narrative account of the scheme is provided in A. Ian Dunlop (ed.) (1992).

the Holy Scriptures as the supreme authority in all matters of faith and conduct, asserting that the Bible is to be interpreted in light of itself. The WCF outlines the doctrine of the Trinity, affirming the Godhead consists of the Father, the Son, and the Holy Spirit. Salvation is another critical theme in the WCF, where it articulates the doctrines of predestination and election, asserting that God has predestined some to eternal life, while others are left to their just condemnation for sin. What does all of this have to do with life annuities? Again, members of the scheme had to accept the WCF to join the pool, and I'll return to matters of predestination and how this might have affected the underlying probability calculations in Chapter 9.

Note that the WCF also details the effects of the fall on humanity, the role of Christ as mediator, and the process of salvation through faith alone. The sacraments of Baptism and the Lord's Supper are defined as means of grace, instituted by Jesus Christ. The WCF also addresses church government, advocating a Presbyterian structure, and discusses civil government's role, emphasizing the importance of moral law.

To summarize, and once again to see the link with Protestant Christianity, the WCF has had a profound influence on Presbyterian churches worldwide, serving as a doctrinal standard and a resource for understanding Reformed theology. Its enduring legacy is seen in its continued use and respect among Reformed Christians as a theological benchmark. Now, I should note that university professors, such as those at Glasgow and Edinburgh, would have been required to swear allegiance to the WCF *independently* of whether they joined the annuity scheme, but it was included as an explicit condition for joining what was the first successful longevity risk-sharing plan.[11]

The First Mandated Savings Plan

Arguably, the annuity scheme was the first-ever mandated savings plan established by an Act of Parliament, forcing all "employees" to contribute a fraction of their wages towards a collective annuity fund.[12] Hew Scott's

[11] For the entire text of WCF describing the main tenets of (seventeenth-century) Calvinist theology see, https://www.apuritansmind.com/westminster-standards/.

[12] This statement is made without prejudice to the earlier Royal Navy's Chatham Chest (1590) "retirement fund" and the occupational pay-as-you-go scheme created by James, Duke of York in 1672. For a chronology of early pension plans, see Lewin (2003).

Fasti Ecclesiae Scoticanae described it as "the first institution of its kind in Scotland."[13] Alexander Webster's obituary in *The Scots Magazine* noted that the church was indebted to him for "the promotion and prosperity of its fund for ministers' widows and children."[14] The renowned eighteenth-century (Welsh) actuary Richard Price acknowledged Webster as "the ingenious founder of the scheme," and the economist Adam Smith (b. 1723, d. 1790)—who actually participated and was a member in the plan—noted that "of all the men I have ever known, Alexander Webster is the most skilful in Political Arithmetic."[15]

This (again) is why modern pension actuaries, and especially if they are Scottish, treat the year 1744 with the same reverence as the year 1693, when Edmond Halley (b. 1656, d. 1742) published the first (in English) valuation of life annuities, or 1725 when Abraham de Moivre (b. 1667, d. 1754) published his book *Annuities Upon Lives*. As one example of the esteem with which 1744 is held, in a late twentieth-century monograph discussing the scheme, the managing director of Standard Life Insurance Company in Edinburgh noted that the 1744 fund "can lay claim to being the first to be established on what are known today as actuarial principles."[16]

Beyond insurance specialists, the famous historians Yuval Noah Harari and Niall Ferguson, writing separately, denote the year 1744 as a human milestone in the transition from religious superstition to scientific modernity. In *Sapiens*, Harari (2014) writes that in seeking guidance on how to design the scheme, Webster and Wallace "did not pray to God to reveal the answer, nor did they search for an answer in the Holy Scriptures or among the works of ancient theologians"; instead, they consulted with Colin Maclaurin to establish the fund on a sound statistical basis. Ferguson (2008), who devotes an entire chapter to the 1744 scheme in his *The Ascent of Money*, breathtakingly remarks that "modern actuaries still marvel at the precision with which [Webster and Wallace] did

[13] Scott (1915, vol. 1, p. 119).

[14] Anonymous (1784), 'Obituary: Sketch of the Character of Dr. Webster', p. 61.

[15] Price quote is in Dunlop (1992), p. 19, and Smith quote is in Michael Anderson. (2011). Guesses, Estimates and Adjustments: Webster's 1755 "Census" of Scotland Revisited Again. *Journal of Scottish Historical Studies, 31*(1), 26–45, 44.

[16] See Dunlop (1992, p. 73).

their calculations."[17] Ferguson makes the startling claim that computations made by Webster and Wallace were so remarkably accurate that their forecast for the projected future value of the fund two decades later was off by only *one pound*. Harari offers this pound as evidence that scientific projections based on probability and statistics were "better than the prophecies of Habakkuk, Jeremiah or St John."[18] Beyond hyperbole, both Harari and Ferguson confuse the 1744 Church of Scotland Scheme for Widows and Orphans of Ministers and Professors with the completely unrelated insurance company Scottish Widows, founded in 1815 to care for widows of soldiers killed in the Napoleonic Wars. All this (noise) needs further examination and underscores the importance of returning to original ecclesiastic sources.

Harari and Ferguson—and other cheerleading chroniclers—seem to suggest that science in the eighteenth century marked a sudden secularizing step towards modernization via the rejection of religious beliefs. And yet, academic historians as well as philosophers such as Peter Harrison, Dominic Erdozain, and J. C. D. Clark, as well as Charles Taylor, unanimously denounce this teleological narrative. By the early twenty-first century, it is widely accepted that many of the well-known Scottish Enlightenment thinkers themselves were inspired by religious ideas.[19]

Lorraine Daston and Ian Hacking have separately emphasized that religious beliefs strongly influenced probabilistic and statistical thinking during the time of the Scottish Enlightenment. Similarly, the intellectual historian Thomas Ahnert has noted the importance of moral culture and argues that it was influenced by a theological language and conception of the Presbyterianism.[20] The point here is that statistical science alone did not, in a sudden big bang of innovation in 1744, solve the long-standing challenge of widows and orphans of ministers.

In particular, I note here the omissions made by so-called popular historical accounts of the 1744 genesis. First, they erroneously—or

[17] Ferguson (2008, p. 193).

[18] Harari (2014, chap. 14), and Ferguson (2008, chap. 4).

[19] For this line of reasoning, see Taylor (2009). On a related theme, see Harrison (2017), Erdozain (2017), Coleman (2010), Clark (2012), Wood (2019), Broadie (2007), and C. Smith (eds).

[20] This is the exact language and wording of University of Edinburgh Professor Thomas Ahnert, and his 2015 book entitled *The Moral Culture of the Scottish Enlightenment: 1690–1805*, which was published by Yale University and is cited as Ahnert (2015).

perhaps even deliberately—underplay the role of the Church of Scotland's governance practices, administrative infrastructure, Presbyterian beliefs, and moral culture in the successful launch of the annuity fund. This first argument engages with the recent literature on accountability practices within the Presbyterian Church as an eighteenth-century manifestation of Max Weber's protestant spirit of capitalism. Indeed, critics of the Weber theory, such as Werner Sombart and Christopher Smout, have argued that the absence of capitalistic enterprises in seventeenth-century Calvinist Scotland refutes his famed thesis. The 1744 fund might offer another counterexample.[21]

Second, the popular perception of 1744 ignores the synod-based origins (the synod was the regional church court) of the scheme in the decades prior to the 1740s, and the evolutionary process by which the national version came into existence, or why it emerged within the church after a period of turmoil. In support of that argument, Chapter 8 engages with the literature on the disruptions with the Church of Scotland in the second, third, and fourth decades of the eighteenth century.

Over the next few chapters I will also reference a parallel economic argument that the initial scheme was actuarially flawed—rather than a paragon of scientific precision—because the contribution rates were too low for most participants, as they did not support the benefits promised to ministers' widows. In fact, the scheme had to be revised a few years later (in 1749), with very tight control over the payment of benefits. Perhaps the scheme's Scottish myth has become intertwined with reality.

Actuarial Historiography

The actuarial historian Trevor Sibbett (2014), in a review of the book by Ian Dunlop (1992) about the 1744 scheme, writes that "the project was the first actuarially sound pension fund in the world." The Danish historian of statistics Anders Hald (2003) notes that "the construction of the 1744 fund helped serve as model for many later pension funds."[22] Actuary Christopher Lewin (2003) writes: "The fund survived for a quarter of a millennium, a splendid tribute to prudential Scottish financial

[21] See McKinlay and Mutch (2015), as well the article by G. Marshall (1980).

[22] Anders Hald (b. 1913, d. 2007).

management."[23] Timothy Johnson (2015), a professor of actuarial mathematics at Herriot Watt University, writes that "the fund was the first fund covering insurance liabilities, either in general insurance or in life insurance, to be managed on a mathematical basis." The design of the fund exported to America and was copied by the Presbyterian Ministers Fund of Philadelphia in 1761. Then, the following year, Equitable Life, the oldest public life insurance company in England adopted a similar approach.[24] In sum, there appears to be little doubt among actuaries and within the community of probability and statistics that this was an important milestone in the development of pension funding.

Prima facie the 1744 fund fits within the Scottish Enlightenment, falling squarely within "institutions to articulate the social, economic and polite programs of this new intellectual class," using the parameters established by Emerson (1988). It was developed in Edinburgh during a "creative surge." Many of the individuals named as trustees were literati. The actuarial work was done by Maclaurin. As Richard Sher (2015) has noted, "In no other European country, Protestant or Catholic, was the Enlightenment so closely tied to, and supported by, the established church and national universities as in Scotland during the age of William Robertson."[25]

This leads us back to the question: *Why isn't the annuity scheme within the established historical canon of formal contributions to the Scottish Enlightenment?*[26] One possibility is its heavy emphasis on computation, and that its main visible output is a simple table of numbers. Might it be that applied mathematical calculations long belittled by purists—versus pure theory—be outside the intellectual range? Is there a bias against (too many) numbers?

In his essay "On the Balance of Trade," David Hume writes: "Every man, who has ever reasoned on this subject, has always proved his theory, whatever it was, by facts and calculations, and by an enumeration of all the commodities sent to all foreign kingdoms."[27] His point was

[23] Source: Lewin (2003, p. 361).

[24] Source: Johnson (2015).

[25] See Sher (2015), introduction.

[26] Houston (1994) mentions the scheme in a footnote and suggests that Historical Demography might be an area where Scottish history could further develop.

[27] David Hume online, https://davidhume.org/texts/empl2/bt#5.

the sense of mistrust in numbers. Indeed, this idea—the politicization of numbers in the seventeenth and eighteenth centuries—is pursued by William Deringer (2018), who writes that "Hume felt that Briton's reckless reverence for numbers helped perpetuate dangerous generalizations about political economy." Deringer continues: "If Hume offered the era's most incisive attack on calculations, Dr. Richard Price offered its most ambitious defense."[28]

Methodist versus Presbyterians

Interestingly, somewhere around the year 1765, Methodists at their Wesleyan Conference decided to establish a fund for "worn out preachers," which was to be financed by itinerant preachers. This fund wasn't meant for widows and orphans, and the annual contributions of half a guinea, where much less than those required for the Scottish Presbyterian fund, as I'll describe in the next chapter. According to the historian of Methodism, Clive Norris, who has also studied the financials of the fund, it also offered life insurance benefits to a deceased preacher's family. Overall, it could best be described as a pure pay-as-you-go scheme, which is something I'll address in Chapter 8, in the context of earlier attempts by the Church of Scotland. In the Methodist fund or scheme, 20 active preachers were needed in order to support one retired and inactive preacher. In fact, John Wesley himself was rather ambivalent about the entire scheme, as he believed poverty was a requirement of their calling, which was a far cry from the Scottish Presbyterian attitude. Methodists often emphasized the importance of generosity and the potential dangers of wealth, while the Presbyterian view, rooted in Calvinism, stressed the sovereignty of God over all aspects of life, including wealth, and the responsibility to manage wealth in a way that honours God. And, in contrasting the Methodist scheme, which ultimately was unsuccessful, with the much better known and quite successful Presbyterian one, Norris credits a "team of mathematicians to undertake the actuarial calculations" as well as them being "grounded in comprehensive information about the

[28] Source: Deringer (2018, p. 285).

potential beneficiaries." No mention is made of the possible impact on (or contribution to) the Scottish Enlightenment.[29]

Beyond actuarial science, and whether that field can and should trace it's modern origins to the Scottish Enlightenment, Michael Mepham notes a similar paternity for the field of business accounting:

> It is suggested that the Scottish ascendancy in accounting texts can be considered as part of the achievements of the Scottish Enlightenment and as complementary to the more renowned works in economics, law and philosophy.[30]

To sum up, whether one accepts the development of the world's first successful longevity risk-sharing scheme as a part of the Scottish Enlightenment (or not), one thing is for certain. The genesis took place in the heart of Edinburgh, during the exact same period of time, involving many of the same people, using similar modes of thinking. More importantly, its impact or longevity lasted well beyond the temporal confines of the Scottish Enlightenment. In the words of Ian Dunlop in his introduction to the 1992 edited book about the scheme: "It was at the forefront of the Enlightenment." So, Scottish historians (of the Enlightenment) should take credit for it or someone else will.[31]

Sources and Additional References

Ahnert, T. (2015). *The Moral Culture of the Scottish Enlightenment: 1690–1805*. Yale University Press.

Broadie, A. (2007). *The Scottish Enlightenment*. Birlinn Limited.

Bruneau, O. (2020). Colin Maclaurin (1698–1746): A Newtonian Between Theory and Practice. *British Journal for the History of Mathematics, 35*(1), 52–62.

Clark, J. C. D. (2012). Secularization and Modernization: The Failure of a Grand Narrative. *The Historical Journal, 55*(1), 161–194.

[29] See Norris (2017), which reviews the history of the plan and the financial circumstances around the demise of the fund by the year 1800.

[30] Quote from abstract of Mepham (1988, p. 151).

[31] For a recent review and discussion of the interaction of science and religion during the period, which will also be addressed in Chapter 9, so the recent book by Mills (2024).

Coleman, C. (2010). Resacralizing the World: The Fate of Secularization in Enlightenment Historiography. *The Journal of Modern History, 82*(2), 368–395.

Dawson, J. (2015). *John Knox*. Yale University Press.

Deringer, W. (2018). *Calculated Values: Finance, Politics, and the Quantitative age*. Harvard University Press.

De Romestin, Rev. H., trans. (2016). *Concerning Widows* by Saint Ambrose. Aeterna Press.

Deuchar, D. (1896). Notes on Widows' Funds. *Transactions of the Actuarial Society of Edinburgh, 3*, 198–227.

Dow, J. B. (1975). Early Actuarial Work in 18th Century Scotland. *Transactions of the Faculty of Actuaries, 33*(240), 193–229.

Dunlop, A. I. (Ed.). (1992). *The Scottish Ministers Widows Fund: 1743–1993*. Saint Andrew Press.

Emerson, R. L. (1988). Science and the Origins and Concerns of the Scottish Enlightenment. *History of Science, 26*(4), 333–366.

Erdozain, D. (2017). A Heavenly Poise: Radical Religion and the Making of the Enlightenment. *Intellectual History Review, 27*(1), 71–96.

Ferguson, N. (2008). *Chapter 4: The Return of Risk in The Ascent of Money: A Financial History of the World*. Penguin.

Grabiner, J. V. (2002). Maclaurin and Newton. In C. W. J. Withers & P. Wood (Eds.), *Science and Medicine in the Scottish Enlightenment* (pp. 143–171). Tuckwell Press.

Hald, A. (2003). *A History of Statistics and Probability and Their Application before 1750*. John Wiley & Sons.

Harari, Y. N. (2014). *Sapiens: A Brief History of Humankind*. Random House.

Harrison, P. (2017). Science and Secularization. *Intellectual History Review, 27*(1), 47–70.

Houston, R. A. (1994). Eighteenth-Century Scottish Studies: Out of the Laager? *Scottish Historical Review, 73*(1), 64–81.

Johnson, T. C. (2015). Reciprocity as a Foundation of Financial Economics. *Journal of Business Ethics, 131*(1), 43–67.

Kahl, S. (2005). The Religious Roots of Modern Poverty Policy: Catholic, Lutheran, and Reformed Protestant Traditions Compared. *European Journal of Sociology, 46*(1), 91–126.

Lewin, C. G. (2003). *Pensions and Insurance Before 1800: A Social History*. Tuckwell Press.

Marshall, G. (1980). *Presbyteries and Profits: Calvinism and the Development of Capitalism in Scotland, 1560–1707*. Clarendon Press.

Mepham, M. J. (1988). The Scottish Enlightenment and the Development of Accounting. *The Accounting Historians Journal, 15*(2), 151–176.

Mills, R. J. W. (2024). Religion and the Science of Human Nature in the Scottish Enlightenment. *Palgrave Pivot*. https://doi.org/10.1007/978-3-031-49031-6

McKinlay, A., & Mutch, A. (2015). Accountable Creatures: Scottish Presbyterianism, Accountability and Managerial Capitalism. *Business History, 57*(2), 241–256.

Norris, C. M. (2017). Chapter 3: The Preacher's Fund in *The Financing of John Wesley's Methodism C. 1740–1800*. Oxford University Press.

Sageng, E. (2004). MacLaurin, Colin (1698–1746), Mathematician and Natural Philosopher. *Oxford Dictionary of National Biography*. Last Retrieved 28 November 2021. https://www.oxforddnb.com/view/10.1093/ref:odnb/9780198614128.001.0001/odnb-9780198614128-e-17643

Scott, Hew. (1915). *Fasti Ecclesiae Scoticanae: The Succession of Ministers in the Church of Scotland from the Reformation*. Oliver and Boyd.

Sher, R. B. (2015). *Church and University in the Scottish Enlightenment* (2nd ed.). Edinburgh University Press.

Sibbett, Trevor. (2014). Widows Fund. In Bjørn Sundt and Jeff L. Teugels (Eds.), *Encyclopedia of Actuarial Science*. John Wiley & Sons.

Taylor, C. (2009). *A Secular Age*. Harvard University Press.

Thane, P. (2020). Old Age in European Cultures: A Significant Presence from Antiquity to the Present. *The American Historical Review, 125*(2), 385–395.

Wood, P. (2019). Science in the Scottish Enlightenment. In A. Broadie & C. Smith (Eds.), *The Cambridge Companion to the Scottish Enlightenment* (pp. 90–112). Cambridge University Press.

CHAPTER 5

A Presbyterian Scheme for Ministers

Introduction & Plan

As noted in Chapter 4, authors going as far back as the late nineteenth century have described the mechanics and details of the *Scheme for Providing a Yearly Annuity to Widows of Ministers of the Established Church of Scotland*, which was its first and official name in the original 1743 proposal. I have also mentioned (repeatedly) the best-selling works of Ferguson (2008) and Harari (2014), both of which describe the scheme and both of which likely obtained their information from a comprehensive 1992 book written and edited by Ian Dunlop and titled *The Scottish Ministers Widows Fund: 1743–1993*. That book, which was published over 30 years ago, (naturally) in Edinburgh, remains the single best secondary reference for the scheme and its full history, from its very beginning in the early 1740s (which is the focus on this book) to its end in the 1990s, when it was merged with a much larger (newer) fund. Dunlop,[1] who coincidentally was a minister in the Church of Scotland, had reviewed most of the archival material about the scheme, which is now stored at the National Records of Scotland, and he included images and pictures of some of its key documents. Dunlop's book, which was reviewed by another well-known historian of actuarial science, Trevor Sibbett (1993), also included chapter articles about the scheme written

[1] For an obituary, see Cheyne (2000), 'In Memoriam: The Rev. Alexander Ian Dunlop'.

M. A. Milevsky, *The Religious Roots of Longevity Risk Sharing*,
https://doi.org/10.1007/978-3-031-62403-2_5

by actuaries David Hare, William Scott, and James Dow. My point here is to emphasize that Dunlop's work offers the single most authoritative (secondary source) description of the fund, and the book by C.G. Lewin (2003), which I mentioned earlier as well, is yet another excellent summary of the scheme's features. For readers who truly want to delve into the actuarial minutia, as well as the history of the fund during the nineteenth and twentieth centuries—neither of which I discuss—Dunlop's book is a great place to start. Figure 5.1 is a front piece of the Act itself, which is freely available online.

So, what can I add here or in this chapter?

My plan is first to provide a very high-level summary of the scheme, which would have been known to (and reviewed by) the above-mentioned authors. But in addition, moving on to Chapter 6 and Chapter 7, I will delve into aspects of the fund—by carefully examining the archival

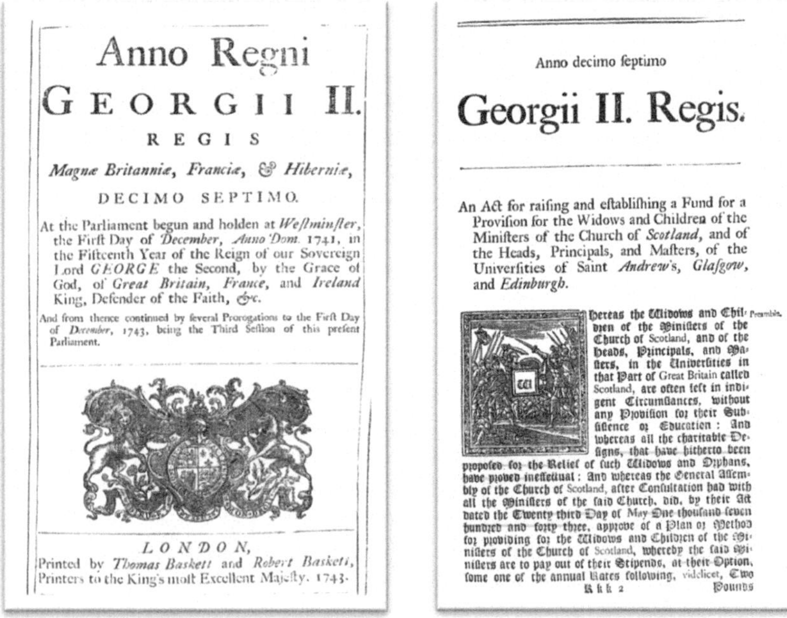
Anno Regni
GEORGII II.
REGIS
Magnæ Britanniæ, Franciæ, & Hiberniæ,
DECIMO SEPTIMO.
At the Parliament begun and holden at *Weſtminſter*, the Firſt Day of *December*, *Anno Dom.* 1741, in the Fifteenth Year of the Reign of our Sovereign Lord *GEORGE* the Second, by the Grace of God, of *Great Britain*, *France*, and *Ireland* King, Defender of the Faith, *&c.*
And from thence continued by ſeveral Prorogations to the Firſt Day of *December*, 1743, being the Third Seſſion of this preſent Parliament.
LONDON,
Printed by *Thomas Baskett* and *Robert Baskett*, Printers to the King's moſt Excellent Majeſty. 1743.

Anno decimo ſeptimo
Georgii II. Regis.
An Act for raiſing and eſtabliſhing a Fund for a Proviſion for the Widows and Children of the Miniſters of the Church of *Scotland*, and of the Heads, Principals, and Maſters, of the Univerſities of Saint *Andrew's*, *Glaſgow*, and *Edinburgh*.
Preamble.
Whereas the Widows and Children of the Miniſters of the Church of Scotland, and of the Heads, Principals, and Maſters, in the Univerſities in that Part of Great Britain called Scotland, are often left in indigent Circumſtances, without any Proviſion for their Subſiſtence or Education: And whereas all the charitable Deſigns, that have hitherto been propoſed for the Relief of ſuch Widows and Orphans, have proved ineffectual: And whereas the General Aſſembly of the Church of Scotland, after Conſultation had with all the Miniſters of the ſaid Church, did, by their Act dated the Twenty third Day of May One thouſand ſeven hundred and forty three, approve of a Plan or Method for providing for the Widows and Children of the Miniſters of the Church of Scotland, whereby the ſaid Miniſters are to pay out of their Stipends, at their Option, ſome one of the annual Rates following, videlicet, Two
Kkk 2 Pounds

Fig. 5.1 The Act of Parliament Taking Effect 25 March 1744 (*Image Source* Great Britain (1743 [1744]) Eighteenth Century Collections Online [ECCO])

records—that were not emphasized or perhaps even unknown to prior generations of writers and historians. In addition to the actuarial and mathematical innovation, which were acknowledged by former authors, I will emphasize the (1) earlier unfunded scheme developed in the 1720s, as well as the (2) Protestant practices and (3) overall managerial aspects that have been neglected in the mathematical and actuarial praise. I will gloss over some aspects, knowing they have been covered in the prior literature and reference them where appropriate, while focusing on other aspects that haven't received as much attention. The aspects on which I opt to shine some extra light are certainly not meant to takeaway or reduce the importance of this milestone in (Scottish) actuarial science. So, while readers who have never heard or read about the 1744 scheme will find sufficient information in the next few pages, my target audience consists of those who are familiar with the scheme and are asking themselves: *What new insight is here*?

High-level Review of the Scheme's Features

At its essence, the annuity scheme was meant to periodically collect, manage, and invest funds to provide workers a *reversionary annuity* for their wives. The money gathered from across the presbyteries and universities in Scotland—and this wasn't easy in the mid-eighteenth century—was invested with the hope or the expectation of earning 4% interest per year. Over time the fund would grow to a sufficient size to pay annuities to an estimated number of widows. More on those actuarial estimates later. But the 4% investment return number was a key financial parameter in the scheme. In some years the fund itself earned more than 4%, and sometimes it earned less than 4% per year, but it was pretty close and within plus or minus 1% per year.[2] I'll explain how that all worked, and why the fund occasionally earned more or less, but for those readers and former authors—such as Ferguson (2008)—who have been "astonished" at how closely the fund value matched its projected value,

[2] I feel obliged at this point in the narrative to emphasize that the 4% target return number for the annuity scheme has absolutely nothing to do with the so-called 4 per cent rule of retirement income planning. The number four is just a coincidence and should not be taken as historical evidence for 4% as a sustainable spending rate in retirement or as an alternative to an annuity. Yes, this might seem obvious to 99.9% of readers, but one never knows these days.

this 4% will be part of the story. To begin, I will simply note that the money (or fund) was invested or allocated in a rather peculiar manner. Instead of "outsourcing" the asset allocation and investment management process to professional advisors (yes, a very anachronistic term), the money was instead invested "in-house." In fact, it was invested *very* in-the-house. How? The contributions (payroll deductions) were collected from contributors once per year and were given back to (a portion of) the contributors in the form of a loan or bond. Yes, the fund managers forced all members *in* the scheme to borrow *from* the scheme. At what rate were they forced to borrow? At the assumed 4% rate underlying the actuarial valuation. No surprise, then, that *realized* returns matched *assumed* returns. Yes, this peculiar asset allocation requires much more elaboration, and not all the funds were allocated in this manner. I'll get to the more granular details in Chapters 6 and 7. For now, let's move on to discuss the annuity payouts.

The benefits were paid in the form of a *reversionary* annuity, which differs from the more familiar and *conventional* life annuity in two key aspects.[3] A conventional annuity has (1) a buyer who receives some benefit during their own lifetime, and (2) a predetermined fixed date or age upon which those benefits commence. These two features are inverted for a reversionary annuity: the buyer never receives any income benefit, since the benefit is instead directed to a secondary person on whose life it depends; and the commencement date of the annuity is not known in advance because the death occurs at a random future time.[4]

Although this might be intuitive or outright obvious to some readers, the economic rationale for buying a reversionary annuity is to generate a secure and adequate income for a spouse (female) upon the demise of the other spouse (male), which would be associated with the sudden termination of a salary or wage. The reversionary annuity is an alternative to basic life insurance which pays a lump sum death benefit. Why wasn't the plan

[3] Once again, I am compelled to remind readers that these are very basic life annuities, and not the modern versions, known as variable annuities with living benefits, fixed-indexed annuities, registered-indexed linked annuities, GLWBs, GMIBs, RILAs, MYGAs, etc. Other than the word annuity in their title, they share little in common with the century's old life annuity.

[4] Reversionary annuities are rarely sold on a stand-alone basis in the twenty-first century. The earliest valuation was by Richard Price in his 1772 article 'Observations on the Proper Method of Calculating the Values of Reversions Depending on Survivorships'. Price will make an appearance later-on. See also Haberman and Sibbett (1995).

or scheme designed as a simple life insurance payout? The rationale was as follows. Inheriting a lump sum—in the words of Alexander Webster—"burdens" the wife with a "stock" that must be invested to generate an annual income. So yes, a design option would have been to create a "life insurance" fund instead of a "reversionary annuity" fund. Alas, this wasn't the approach taken by the scheme designers because they felt it would place an undue burden on the widow to invest, manage, and decumulate the fund to generate income. The life annuity was a better way to achieve this goal, an insight as relevant today as it was in the eighteenth century.

Here it is in Alexander Webster's words: 'The committee cannot help thinking that £20 payable yearly to the mother will better answer the purposes, than the [one lump] sum of £200." Interestingly, this not only sets the life annuity factor at ten-to-one, it is also a rather clever (and modern) insight into human behavioural finance. The scheme plan documents note:

> The advantages of providing a yearly annuity to minister's relicts are obvious and many. And, considering how often fruitless attempts have been made for this purpose, [the scheme] we hope will give universal satisfaction that something is at length hit upon which has already gained the unanimous approbation of the General Assembly.[5]

Recall that the General Assembly of the Church of Scotland was the main administrative body or management committee, which met annually to manage the affairs of the church, and the trustees of the annuity fund—more on them later—would have to report back to the General Assembly annually on the state of affairs. Note also the mention of prior "fruitless attempts" which will be the topic of Chapter 8, where I discuss the earlier attempts by the local synods.[6]

Now, despite the "pension mindset" that has permeated the first few chapters of this book, and in particular the ranking of the national scheme in Chapter 1, as well as the benefits of longevity risk pooling in Chapter 3, I must be crystal clear here about one very important fact. Namely, the 1744 annuity scheme was not an old-age pension plan in any modern sense of the phrase. Ministers and professors in the eighteenth century certainly did not pick some arbitrary date or age (e.g. age 65) and then

[5] CH9/17/2, 'A Scheme for providing ...', p. 6.

[6] It actually continues to do so, to this date. See the Church of Scotland website.

"retire" from their duties and start collecting a pension for life while they spent their days golfing or gardening. That is a very modern invention.

Nevertheless, the reversionary annuity scheme for widows of ministers and professors was clearly "longevity risk pooling" and is only a small intellectual step away from a proper retirement pension plan that provides joint-life annuities at a fixed chronological (retirement) age to both the contributor and the surviving spouse. More importantly, one can trace an intellectual line of sight from Alexander Webster—again, more on him later—to Richard Price, then to Francis Maseres (b. 1731, d. 1824), then to the Marquis de Condorcet (b. 1743, d. 1794), and finally to Otto von Bismarck (b. 1815, d. 1898), who implemented the world's first national pension scheme. I'll get back to those five key names and links in the chain of modern pensions later on in Chapter 11.[7] For now, let's move on to discuss the financial magnitude of benefits (to the widow) and what they cost (the contributor).

A Note on Money

For those (younger) readers who are (only) familiar with the decimal system of money implemented after 1971, British currency in the eighteenth century was reported based on the pound, shillings, and pence system. At that time, 12 pence (d) was equal to one shilling (s) and 20 shillings would equal a pound (£). Think of four quarters equalling a dollar or five nickels equalling a quarter. These are just names and will be written as £X.Y.Z for consistency and ease of reading. In that triplet, the letter X is for pounds, Y for shilling, and Z for pence. And, since there are $12 \times 20 = 240$ pence to a pound, the decimal equivalent of £X.Y.Z is computed via $(12Y + Z)/240$. So, for example, the triple written as £1.2s. 3d would be equivalent to one pound (the first number) plus 27/240 of a pound, or £1.1125 after the date of 15 February 1971, known as Decimal Day. This chapter and most of this book will default to the pre-1971 notation for consistency unless specific computations require decimalization.

[7] See Goetzmann (2016).

The Reversionary Annuity: Choices

In early 1744, when the scheme went into operation, *new* ministers and professors who joined their "employer" were forced to participate in and contribute to a savings plan that was meant to fund the reversionary annuity. They also had some financial choices to make. More on that to come, but in contrast to new participants who joined after 1744, *old* ministers and professors were given the option to opt out of the scheme. I'll get to the exact mechanics of how this was implemented later. At a high level, Fig. 5.2 displays the official form or paperwork that participants would have to complete in order to notify the administrators of the scheme that they wanted to participate and contribute towards funding a reversionary annuity for their wives.

Although that particular printed form in Fig. 5.1 only came into effect later in the eighteenth century, and initially the notifications were done in handwritten notes, the main features or choices were the same. The

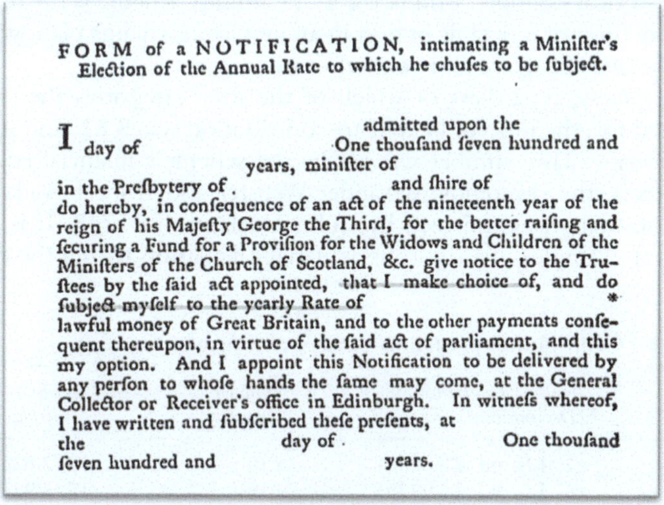

FORM of a NOTIFICATION, intimating a Miniſter's Election of the Annual Rate to which he chuſes to be ſubject.

I admitted upon the day of One thouſand ſeven hundred and years, miniſter of in the Preſbytery of . and ſhire of do hereby, in conſequence of an act of the nineteenth year of the reign of his Majeſty George the Third, for the better raiſing and ſecuring a Fund for a Proviſion for the Widows and Children of the Miniſters of the Church of Scotland, &c. give notice to the Truſtees by the ſaid act appointed, that I make choice of, and do ſubject myſelf to the yearly Rate of * lawful money of Great Britain, and to the other payments conſequent thereupon, in virtue of the ſaid act of parliament, and this my option. And I appoint this Notification to be delivered by any perſon to whoſe hands the ſame may come, at the General Collector or Receiver's office in Edinburgh. In witneſs whereof, I have written and ſubſcribed theſe preſents, at the day of . One thouſand ſeven hundred and years.

Fig. 5.2 Scheme choice notification forms[8]

[8] Source is the National Records of Scotland (NRS), and permission to reproduce was granted.

participant noted his name, his location, the date, and, most importantly, his choice.

What was the big choice to be made? Namely, in the annuity scheme all participants could select a relatively high spousal annuity benefit of £25, in exchange for contributing (think of insurance premiums of) £6 11s. 3d. per annum, or they could opt for a £10 annuity in exchange for £2 12s. 6d. There were two other options, namely, a £15 or a £20 annuity, with distinct contributions or premium rates. But, once these premium rates were selected, they were irreversible for life. The participants couldn't change their minds and scale up or down, ever. All four classes as they were known are listed in Table 5.1

Just as importantly, in the absence of an active election—that is if the ministers or professors didn't bother or forgot to send in their election or forms—they were "defaulted" into the second class, associated with a yearly premium contribution of £3. 18s. 9d. Why was that the default? The justification given for this rate was that it was the medium of the four available options. So, at this point, and in the absence of locating an earlier historical example, I think the 1744 annuity scheme is the first-ever implementation of a default option in an employee savings plan, with full awareness that those are all twentieth-century terms.[9]

Nevertheless, regardless of which of the four categories the contributor opted for, the ratio of benefit-to-contribution was 3.81, and it scaled proportionally. This number was set by the scheme's financial engineers or architects, the reverends Alexander Webster and Robert Wallace in a calculation that was confirmed by Colin Maclaurin in 1743. It is one of the earliest examples of a rigorous actuarial calculation within the context

Table 5.1 Contribution premiums versus benefit payouts

Class	*Annual Contribution (Decimalized)*	*Widows' Annuity*	*Payout Multiple*
#1	£2. 12s. 6d (£2.6250)	£10	10/2.625 = 3.81
#2	£3. 18s. 9d (£3.9375)	£15	15/3.937 = 3.81
#3	£5. 5s. 0d (£5.2500)	£20	20/5.250 = 3.81
#4	£6. 11s. 3d (£6.5625)	£25	20/6.562 = 3.81

Source Great Britain (1744, 1749), and Assembly Acts (1742)

[9] I elaborated on this in the article cited as Milevsky (2023).

of pension annuities, and Dunlop (1992) makes that point quite clearly. Indeed, there was a lot of mathematical work and perhaps "Newtonian calculus" that went into deriving that single 3.81 number, which I'll try to explain in Chapter 7, but Scottish actuaries should be very proud of this (enlightened) achievement.[10]

Mathematically, though, while the above-noted hard-learned multiple of 3.81, or ratio of contribution to benefit, might be appropriate for a representative individual who is (say) x-years old and whose wife is (say) y-years old, mixing all participants of different ages together and offering them all the *exact same multiple* leads to some inequities. Intuitively, older people (say a 60-year-old minister) should be paying more for a reversionary annuity than a younger person (say a 30-year-old minister), all else being equal. After all, the senior won't be paying premiums for very long, yet the hypothetical benefit is the same. More importantly, if the older participants had a choice of which benefit to select, it's quite natural to expect them to skew towards the large annuities, if the price is the same. Insurance economists call this anti-selection. Stated differently, the fixed 3.81 multiple effectively undercharges older people, who were subsidized by younger members. Now, this focus on insurance, economics, and actuarial minutia might sound rather arrogant from the computational perspective of the twenty-first century, but it should not distract from the administrative and welfare achievements of the fund.[11]

Nevertheless, this pricing flaw is relevant to the "big bang myth" propagated by popular historians that this scheme emerged in 1744 as a perfect plan, whole cloth. Indeed, not only was the 1744 scheme an outgrowth of the earlier (failed) synod-based schemes, which I'll discuss in Chapter 8, this actuarial problem led to financial problems that escalated over time, because the mispricing was exacerbated by choices. Let's look at those choices.

[10] Lewin (2003), writes 'a splendid tribute to prudential Scottish financial management', pp. 361–367. See also the actuarial analysis by Hare and Scott (1992) as well as, more recently, Allen (2022). calculations are part of CH9/17/16 and produced in Ferguson (2008, p. 195).

[11] Chapter 7 provides some additional actuarial context for the reversionary annuity, as does Milevsky (2023). See also Dunlop (1992), and specifically Chapter 2 by J.B. Dow and Chapter 3 by D. J. P. Hare for a full actuarial analysis, p. 83 contains a photograph of Colin Maclaurin's calculations (of the 3.81 multiple) which are dated to May 23, 1743.

Choice Architecture versus Choices Made

When the plan was originally drafted, in late 1743, the scheme architects had to make projections regarding how many of the ministers might select the expensive £25 annuity versus the cheaper £10 one. This initial budgeting estimate was very widely off the mark. In particular, few ministers or professors selected the default option (class #2, £15 annuity). Table 5.2 provides a hand-collected summary of actual versus assumed choices made by close to 1,250 participants noted in the Trustees' Scheme Books. For example, the architects assumed that 54% of ministers and professors would end up in the second-class default option. Perhaps they also viewed this class as the most appealing choice, and that is why they made it the default option. In the scheme's documents, they did not project how many would actively select this class versus defaulting into that category. But in practice, only 8.8% (77 out of 873) of incumbents actively selected the second class, and a further 12.6% (110 out of 873) of incumbents failed to respond and were therefore "defaulted." The point is, they were widely off the mark.

When the dust settled on the initial year of the fund, and everyone either made their choices or were defaulted, the majority of the 873 incumbent ministers and professors selected the higher, more expensive, and more lucrative third and fourth classes, or annuities of £20 and £25. Also, the behaviour of ministers differed from professors, and the former were more likely to not choose (i.e., do nothing, ignore) and thus were defaulted into the second option. The professors were more active, with

Table 5.2 Scheme choice assumptions versus realizations

	Assumed selection (%)	*Actual selection (%)*	*Deaths (#) 1744–1755*	*Deaths (%) 1744–1755*
Class #1 (£10)	21.5	3.6	10	3
Class #2 (£15)	53.8	21.4% =	50 =	18% =
Active & Default		8.8% + 12.6	16 + 34	6% + 12
Class #3 (£20)	21.5	33.3	92	32
Class #4 (£25)	3.2	26.8	79	27
Opted Out	0.0	14.9	57	20
Total:	100	100	288	100

Source Column 1, CH9/17/1, "Representation and scheme...,"Columns 2, 3, 4, CH9/4/2, "Trustees' Scheme Books, 1744–1755."

many writing letters to declare and demand that they not be placed in any category, and thus opted out.[12]

Here is the main takeaway and a precursor to some of the later financial problems experienced by the scheme. The largest gap between the plan's assumptions and reality was between the assumption that 3.25 (in the scheme documents) would select the highest most expensive fourth class of £25, while in practice a much larger 26.8% actively selected that class.[31] Now, the large migration towards the most lucrative benefits might have created a (false) narrative that older ministers and professors were more likely to select the higher benefits. In fact everyone gravitated towards the higher benefits. Why? Perhaps it seemed like a good plan and all participants wanted the highest benefits.

Columns 2, 3, 4, CH9/4/2, "Trustees' Scheme Books, 1744–1755".

Once again, the most important takeaway from Table 5.2 is that designers or architects underestimated the number who would opt for the highest benefit. This is critical because the scheme "underpriced" (a 3.1 multiple, when it should be higher) reversionary annuities for most participants. The skew towards more expensive (£25) benefits threatened the viability of the fund. Furthermore, to exacerbate matters and increase the financial pressure on the fund, 59% of deaths of ministers and professors—which triggered a payout, see the last column of Table 5.2—during the first decade of the scheme's existence, were participants in the top two classes.[13]

A Glimpse into the Archives

To give readers an intuitive sense of what these archival documents look like, Fig. 5.3 is but one of the hundreds of such pages of documents, records, and lists in the National Records of Scotland.[14] That page displays the handwritten names, choices, contributions, and payments of professors at the University of Glasgow. Every presbytery and university had a similar page and list. Readers who might have (and are interested in) Scottish ancestry, or those versed in the historiography around Scottish

[12] Chapter 6 will discuss some of those individual choices.

[13] This, again, was hand collected from the Scheme Books, and in particular CH9/4, CH9/5 and CH9/6.

[14] Reproduced with permission and thanks. Ch9/7/1 (p. 35).

universities in the mid-eighteenth century, are likely to recognize many of the names on the list. In particular note the name of Robert Simpson, who was a Professor of Mathematics at the University of Glasgow. His name is the fourth on the list. Then look across to the middle column and the letter D, which means he sent in his "declinature" and refused to participate in the scheme. Like others with the letter D, he didn't want to contribute or participate in the scheme. Perhaps he couldn't afford to, or perhaps there were deeper reasons to stay away from a plan administered by the Church of Scotland. More on the decliners (and perhaps why) in Chapter 6, which includes mention of key individuals and members of the Scottish Enlightenment.

35

Scheme No. V. Relating to the Charge on the General Collector

A General List of all Ministers in the Church & all Professors in the several Univers.

University of Glasgow

Professors Names	Names of their respective Offices	Dates of their Admission	Notifying	Declinature			Dates of the Writings
1 Mr. Neil Campbel	Principal	1728	Notifying				15 April 1744
2 Mr. John Loudoun	professor of Logick	25 May 1699		Declinature			18 Septr 1744
3 Mr. Alexr. Dunlop Senior	professor of Greek	26 Septr 1704	Do				15 August 1744
4 Mr. Robert Simpson	Professor of Mathematicks	20 Novr 1711		Do			15 Augt 1744
5 Mr. William Forbes	professor of Civil Law	1714		Do			18 July 1744
6 Doctor John Johnstoun	professor of Medicine	June 1714		Do			14 Septr 1744
7 Mr. Robert Dick	professor of Natural philosophy	Octr 1714	Do				27 April 1744
8 Mr. William Anderson	professor of Church History	August 1721	Do				25 May 1744
9 Mr. Francis Hutcheson	professor of Moral philosophy	3 Novr 1730		Do			3 Septr 1744
10 Mr. George Ross	professor of Humanity	11 Decr 1735	Do				1744
11 Mr. Robert Hamilton	professor of Anatomy & Botany	June 1742		Do			14 Septr 1744
12 Mr. William Leechman	professor of Divinity	1744	Do				May 1744
13 Mr. Alexr. Dunlop Junior	professor of Oriental Languages	23 Octr 1744	Do				

Fig. 5.3 Glasgow professors and 1744 scheme lists

Where Are These Financial Contributions Material?

To contextualize the financial burden these contributions might have placed on a typical minister's household budget in the eighteenth century, I'll note that the minimum stipend within the Church of Scotland was fixed (in 1633) at £45 per year, paid in either "money or victual." In fact, a separate church commission to consider the augmentation of ministers' stipends, struck in November 1749, reported that out of 1,000 or so parish ministers, 16 received a stipend of less than £35 per year, 41 received less than £40, and 147 received the legal minimum of £45. The remainder earned more than the legal minimum. The same commission estimated that necessary annual expenditures of a minister living south of the River Forth were at least £83 per year.[15] As yet another calibration point, Richard Sher notes stipends of £139 for Rev. Hugh Blair, £94 for Rev. Alexander Carlyle, and £75 for Rev. William Robertson.[16]

My argument here is that contribution rates of between 5 and 15% of the minimum stipends placed a relatively high financial burden on many ministers, which helps explain why participation was optional, but only for incumbents.[17] This also helps place in proper context why soon after the scheme was amended in late 1748—to help put the fund on a better financial footing—proposals to augment stipends for all ministers were placed before the General Assembly.[18] Further connecting the widows' scheme to church stipends, in their discussion of the patronage struggles, historians of the period, Richard Sher and Alexander Murdoch note the distractions that came from (1) agitation to augment minister stipends, and (2) the "forty-five"; implying both events were on the same order of importance.[19] It seems the creation of a 1744 fund to help widows

[15] Sefton (1962, p. 110); Morren (1838, p. 191).

[16] Sher (2015, p. 122).

[17] Although perhaps contributions for the Moderates didn't pose as big a financial burden.

[18] The amended plan is filed as: *An Act for explaining and amending an Act passed in the Seventeenth Year of His present Majesty's Reign, intituled [sic], An Act for raising and establishing a Fund, for a Provision for the Widows and Children of the Ministers of the Church of Scotland, and of the Heads, Principals, and Masters, of the Universities of Saint Andrews, Glasgow, and Edinburgh.* 22 Geo. II, c. 21, and received royal assent on 26 May 1749.

[19] Sher and Murdoch (1983, p. 274).

and orphans of deceased ministers might have had the opposite effect. Instead of relieving financial stress (for the widows), it increased the constraints on the husbands. This tension between individual members' financial interests hints at a key historical argument related to the role of the church and religion in the scheme. Namely, rather than the annuity scheme being positioned as *exogenously* appended to—or perhaps separate from—the history of the Church of Scotland, its genesis might have also endogenously impacted ecclesiastical church matters. It also speaks to the connection between longevity risk pooling and the religious roots of the scheme.

Revisions to the Founding Text

Now, because the 3.81 multiple was (effectively) too low, and because so many of those who actively selected to participate chose the highest benefit available, the parameters of the scheme had to be changed. The process and chronology by which that took place is explained and laid out in Chapter 6 and Chapter 7. Here, I'll simply note that in his attempt to justify the required amendments to the scheme, Reverend Alexander Webster wrote the following in the lead-up to the (revised) Act of 1749:

> The provisions of such ministers as should die soonest, even in the first year of the scheme, [who] consequently contribute the least were to be as great as the provisions of those who should live the longest and contribute most, which…[would] be much too favorable for those of advanced age, were however generously agreed to by the younger brethren. It always be understood that if the fund should not answer, application be made to the legislature for such alterations.[20]

Note here the phrase "were however generously agreed to by the younger brethren." The younger (newer) participants were aware of this economic subsidy or financial transfer, but were still willing to be part of the same large pool. Remember the context of a church-run scheme, in which the younger brethren care for the interests of the older ones. I'll return to

[20] CH9/17/34, 'Calculations with the Principles and Data on which They Are Instituted Relative to a Late Act of Parliament, Showing the Rise and Progress of the Fund'.

this exact point—and whether it applies to longevity risk-sharing pools in the twenty-first century—in the final chapter of this book.

Nevertheless, here is one example of this favourable advantage. The first minister (or his wife) to benefit from the scheme was James Anderson from Roseneath in Dumbarton Presbytery. He died on 28 June 1744, three months after the launch of the scheme. Anderson had fortuitously (like most others) selected the highest benefit and likely contributed no more than one year's rate of £6 11s. 3d. when he died in June 1744, but his wife, Margaret Turner, enjoyed a benefit of £25 per annum for the rest of her life.[21]

Additional generosities placed further financial pressures on the scheme, such as this important and key element. In the absence of a surviving widow, children under the age of 16 were entitled to a lump sum benefit (called "a stock") of ten times what her annuity benefit might have been. And, even if a widow was still alive at the time of her husband's death, if she died within 10 years, the children would be entitled to the remaining difference as a lump sum. Other aspects of the scheme that were linked to "church politics" (for lack of a better term), were that (1) every vacant position was subjected to an extra tax of approximately £3 per year, (2) ministers and professors who married had to double up on their contributions for the year in which they married, (3) and the annuity payment would be extinguished if the beneficiary ceased being a widow.[22] These all must have affected church life, and not all in a positive way.

In fact, the termination of benefits upon remarriage created a financial impediment to remarriage, and in numerous cases that reached scheme's trustees and required legal action. Namely, widows remarried but failed to disclose their new status to the trustees. The annual reports make for interesting reading, and in one titillating case a widow was caught and then denied being married despite the existence of a 25-year-old marriage licence. The trustees demanded that hundreds of pounds be returned, and in response, she requested that her 25-year-old marriage be annulled; she eventually lost the case.[23]

[21] CH9/6/2, 'Accounts, Cash Received and Paid, 1744–1755' (Author unable to locate Turner's date of death).

[22] See Great Britain (1744). It received Royal Assent as 17 Geo. II, c.11 on 2 March 1744. See also, Great Britain (1749), which codified the 1749 revisions.

[23] CH9/16/2, 'Printed Reports', 1747–1790.

A Protestant Management Ethic

This anecdote, which talks to the micro-management of affairs, captures what I would like to argue is a key innovation in the scheme. Namely, the improvement wasn't the discovery of the above-noted 3.81 contribution (actuarial) multiple. In theory, it is relatively easy to design a plan to which participants make annual contributions, money is collected, funds are invested and benefits are paid to widows and orphans: this is as true in the twenty-first century as it was in the eighteenth. The challenge is the implementation or creation of oversight and management control. This is precisely where the bureaucracy of the Church of Scotland shines bright and is perhaps the source of the scheme's longevity. Presbyterian ministers put in place an infrastructure to combat fraud, keep ample cross-validated records, and continuously monitor life and death. The source documents contain multiple and repetitive lists of contributors and their entitlements, the number of widows and how long they have been alive, the names and ages of all children, dates, payments, exact sums, and who owed what and when. When viewed in totality, the impression is that the administrative work this scheme entailed was simply breathtaking.

The trustees, clerks, and collectors repeatedly implored the synods, presbyteries, and parishes to keep better records, send in their lists of contributors and beneficiaries on time, record them in a particular manner, use specific formatting, and delete empty spaces. Financial fines, heftier than the interest on tardy contributions, were also imposed on delinquent administrators. The Assembly Acts provide one example, from May 1747:

> Pertaining to the fund for widows and orphans...we enjoin the presbyteries...hereby appointed and ordained that the annual lists to be transmitted from time to time to the trustees shall be drawn up in the precise words, columns and order contained the specimens that have been sent to the several presbyteries. And in the case the presbyteries shall fail in making their annual returns at the time and the manner foresaid the assembly do recommend to the trustees to exert those powers the law has put into their hands.[24]

[24] The Acts of the General Assembly of the Church of Scotland, Edinburgh, 15 May 1747, p. 7.

Ten years later, in May 1757, an Act[25] was passed to ensure that "separate registers shall be divided into many parts as the respective presbyteries consist of parishes," and that every presbytery and university should ensure their lists are "bound up with a quire of clean paper." The administration adapted, but the ratio of 3.81 never changed.[26]

The scheme owes part of its historical significance to the adaptive bureaucracy it imposed on members. And, echoing Max Weber's spirit of capitalism alluded to above,[27] perhaps only a strict Presbyterian church could impose the discipline and self-reflection required to operate the scheme. To be clear, Margaret Jacob and Matthew Kadane,[28] and most modern sociologists and economists, reject the mass "Weberian" idea that Protestant beliefs led to capitalism or that Max Weber even made that claim.[29]

Rather, the point here is that T. C. Smout, in the *History of Scottish People, 1560–1830*, and other critics cite the example of Scotland as a refutation of Max Weber's spirit of capitalism. But they may have missed the 1744 scheme and the reflective and process-oriented leitmotif.[30] This is also consistent with current views of historians such as Alan McKinlay and Alistair Mutch.[31] They build on the work of Gordon Marshall,[32] and argue that separate from Protestant *beliefs*, "systematic routine accountability practices" evolving from Protestant *practices* nudged the development of managerial capitalism and financial accountability. The scheme, when viewed in totality, offers a Scottish example of Weber's spirit, resonating with Marshall's claim that "the history of the development of capitalism in Scotland may be consistent with Weber's argument."[33] In sum, the 1744 fund might not have been a triumph of

[25] This appears immediately after an admonition against attending theatre, as John Home's play *Douglas* had premiered five months earlier, on 15 December 1756.

[26] Church of Scotland, General Assembly (1743), *A Scheme, for Providing an Annuity to Ministers Widows* ..., p. 7.

[27] See Weber ([1905] 2002) as well as the interpretation by Tawney ([1926] 2021).

[28] Jacob and Kadane (2003).

[29] See Ghosh (2019), as well as Becker and Woessmann (2009).

[30] Smout (1972). See also Marshall (1980a), and Hudson and Coukos (2005).

[31] McKinlay and Mutch (2015), and Mutch (2017).

[32] Marshall (1980b).

[33] Marshall (1980a, p. 419).

actuarial science over superstition during the Scottish Enlightenment as much as an important contribution to managerial accounting.[34]

Sources and Additional References

Primary

Church of Scotland. General, Assembly. (1743). *A Scheme, for Providing an Annuity to Ministers Widows, and a Stock to Their Children; Humbly Offered to the Consideration of the Ministers of the Established Church of Scotland, by a Committee of the Commission of the Late General Assembly, Appointed to Draw up the Same, and Transmit It to the Several Presbyteries*. Printed by R. Fleming and A. Alison. Eighteenth Century Collections Online, https://link.gale.com/apps/doc/CB0126539754/ECCO?u=ed_itw&sid=bookmark-ECCO&xid=fcf599e7&pg=1. Accessed 12 July 2022.

Great Britain. (1744). *An Act for Raising and Establishing a Fund for a Provision for the Widows and Children of the Ministers of the Church of Scotland, and of the Heads, Principals, and Masters, of the Universities of Saint Andrew's, Glasgow, and Edinburgh*. Thomas Baskett and Robert Baskett. Eighteenth Century Collections Online, http://link.gale.com/apps/doc/CB0126931101/ECCO?u=ed_itw&sid=bookmark-ECCO&xid=5f48cfc9&pg=1. Accessed 12 July 2022.

Great Britain. (1749). *An Act for explaining and amending an Act passed in the seventeenth year of His present Majesty's reign, intituled, An Act for raising and establishing a fund for a provision for the widows and children of the ministers of the Church of Scotland, and of the heads, principals, and masters of the Universities of Saint Andrews, Glasgow, and Edinburgh*. Thomas Baskett and Robert Baskett. Eighteenth Century Collections Online, https://link.gale.com/apps/doc/CB0126931184/ECCO?u=ed_itw&sid=bookmark-ECCO&xid=5fb0b1ea&pg=1. Accessed 18 July 2022.

Scott, Hew. (1915). *Fasti Ecclesiae Scoticanae: The Succession of Ministers in the Church of Scotland from the Reformation*. Oliver and Boyd.

[34] Emerson (1988), and see also Mepham (1988).

National Records of Scotland (NRS) Church of Scotland (CH) Records

CH9/1: Church of Scotland Trustee Minutes; NRS Catalogue—catalogue record (nrscotland.gov.uk). https://catalogue.nrscotland.gov.uk/nrsonlinecatalogue/details.aspx?reference=CH9%2f1%2f13&st=1&tc=y&tl=n&tn=n&tp=n&k=&ko=a&r=ch9%2f1&ro=m&df=&dt=&di=y

CH9/3: Reports; NRS Catalogue—search results overview (nrscotland.gov.uk). https://catalogue.nrscotland.gov.uk/nrsonlinecatalogue/overview.aspx?st=1&tc=y&tl=n&tn=n&tp=n&k=&ko=a&r=ch9%2f3&ro=m&df=&dt=&di=y

CH9/4: Trustees' Scheme Books; NRS Catalogue—search results overview (nrscotland.gov.uk). https://catalogue.nrscotland.gov.uk/nrsonlinecatalogue/overview.aspx?st=1&tc=y&tl=n&tn=n&tp=n&k=&ko=a&r=CH9%2f4&ro=m&df=&dt=&di=y

- CH9/4/2: "Trustees' Scheme Books, 1744–1755."

CH9/5: Cash Books; NRS Catalogue—search results overview (nrscotland.gov.uk). https://catalogue.nrscotland.gov.uk/nrsonlinecatalogue/overview.aspx?st=1&tc=y&tl=n&tn=n&tp=n&k=&ko=a&r=CH9%2f4&ro=m&df=&dt=&di=y

CH9/6: Accounts, Cash Received and Paid; NRS Catalogue—search results overview (nrscotland.gov.uk). https://catalogue.nrscotland.gov.uk/nrsonlinecatalogue/overview.aspx?st=1&tc=y&tl=n&tn=n&tp=n&k=&ko=a&r=CH9%2f4&ro=m&df=&dt=&di=y

- CH9/6/2: "Accounts, Cash Received and Paid, 1747–1748."

CH9/7: Clerk's Ledgers; NRS Catalogue—search results overview (nrscotland.gov.uk). https://catalogue.nrscotland.gov.uk/nrsonlinecatalogue/overview.aspx?st=1&tc=y&tl=n&tn=n&tp=n&k=&ko=a&r=CH9%2f4&ro=m&df=&dt=&di=y

CH9/10: Schemes Re Charge of General Collector; NRS Catalogue—search results overview (nrscotland.gov.uk). https://catalogue.nrscotland.gov.uk/nrsonlinecatalogue/overview.aspx?st=1&tc=y&tl=n&tn=n&tp=n&k=&ko=a&r=CH9%2f4&ro=m&df=&dt=&di=y

CH9/12: Abstract of Facts and Remarks; NRS Catalogue—search results overview (nrscotland.gov.uk). https://catalogue.nrscotland.gov.uk/nrsonlinecatalogue/overview.aspx?st=1&tc=y&tl=n&tn=n&tp=n&k=&ko=a&r=CH9%2f4&ro=m&df=&dt=&di=y

CH9/15: Miscellaneous Volumes; NRS Catalogue—search results overview (nrscotland.gov.uk). https://catalogue.nrscotland.gov.uk/nrsonlinecatalogue/overview.aspx?st=1&tc=y&tl=n&tn=n&tp=n&k=&ko=a&r=CH9%2f4&ro=m&df=&dt=&di=y

CH9/16: Printed Reports; NRS Catalogue—search results overview (nrscotland.gov.uk). https://catalogue.nrscotland.gov.uk/nrsonlinecatalogue/overview.aspx?st=1&tc=y&tl=n&tn=n&tp=n&k=&ko=a&r=CH9%2f4&ro=m&df=&dt=&di=y

- CH9/16/1: "Printed Reports, 1744–1750."
- CH9/16/2: "Printed Reports, 1747–1790."
- CH9/16/3: "Printed Reports, 1751–1760".

CH9/17: Miscellaneous Documents; NRS Catalogue—search results overview (nrscotland.gov.uk). https://catalogue.nrscotland.gov.uk/nrsonlinecatalogue/overview.aspx?st=1&tc=y&tl=n&tn=n&tp=n&k=&ko=a&r=CH9%2f4&ro=m&df=&dt=&di=y

- CH9/17/2: "Scheme for providing an annuity to ministers widows, and a stock to their children; passed and resolved by the General Assembly ... 12 May 1743." R. Fleming & Co., Edinburgh.
- CH9/17/34: "Calculations, with the principles and data on which they are instituted ... shewing the rise and progress of the fund." T. Lumisden & Co., Edinburgh, 1748.

Secondary

Allen, A. (2022). Corporate Charity for "the House": Craft Pensions and the Widows' Fund, 1670–1782. *Scottish Historical Review, 101*, 210–247.

Becker, S., & Woessmann, L. (2009). Was Weber Wrong? A Human Capital Theory of Protestant Economic History. *The Quarterly Journal of Economics, 124*(2), 531–596.

Becker, S. O., Rubin, J., & Woessmann, L. (2021). Religion in Economic History: A Survey. *The Handbook of Historical Economics*, 585–639.

Cheyne, A. C. (2000). In Memoriam: The Rev. Alexander Ian Dunlop. *Scottish Church History, 30*(1), vii–viii.

Dunlop, A. I. (Ed.). (1992). *The Scottish Ministers Widows Fund: 1743–1993.* Saint Andrew Press.

Emerson, R. L. (1988). Science and the Origins and Concerns of the Scottish Enlightenment. *History of Science, 26*(4), 333–366.

Ferguson, N. (2008). *Chapter 4: The Return of Risk in The Ascent of Money: A Financial History of the World*. Penguin.

Ghosh, P. (2019). History and Theory in Max Weber's Protestant Ethic. *Global Intellectual History, 4*(2), 121–155.

Goetzmann, W. (2016). *Money Changes Everything: How Finance Made Civilization Possible*. Princeton University Press.

Great Britain. (1744). *An Act for Raising and Establishing a Fund for a Provision for the Widows and Children of the Ministers of the Church of Scotland, and of the Heads, Principals, and Masters, of the Universities of Saint Andrew's, Glasgow, and Edinburgh*. Thomas Baskett and Robert Baskett. Eighteenth Century Collections Online. Accessed 12 July 2022.

Great Britain. (1749). *An Act for explaining and amending an Act passed in the seventeenth year of His present Majesty's reign, intituled, An Act for raising and establishing a fund for a provision for the widows and children of the ministers of the Church of Scotland, and of the heads, principals, and masters of the Universities of Saint Andrews, Glasgow, and Edinburgh*. Thomas Baskett and Robert Baskett. Eighteenth Century Collections Online. Accessed 18 July 2022.

Haberman, S., & Sibbett, T. A. (Eds.). (1995). *History of Actuarial Science*, 10 Volumes. Published by William Pickering.

Harari, Y. N. (2014). *Sapiens: A Brief History of Humankind*. Random House.

Hare, J. P., & Scott, W. F. (1992). 'The Scottish Ministers' Widows' Fund of 1744. In A. Ian Dunlop (Ed.), *The Scottish Ministers' Widows' Fund 1743–1993* (pp. 23–55). Saint Andrew Press.

Hudson, K., & Coukos, A. (2005). The Dark Side of the Protestant Ethic. *Sociological Theory, 23*(1), 1–24.

Jacob, M. C., & Kadane, M. (2003). Missing, Now Found in the Eighteenth Century: Weber's Protestant Capitalist. *The American Historical Review, 108*(1), 20–49.

Lewin, C. G. (2003). *Pensions and Insurance before 1800: A Social History*. Tuckwell Press.

Marshall, G. (1980a). The Dark Side of the Weber Thesis: The Case of Scotland. *The British Journal of Sociology, 31*(3), 419–440.

Marshall, G. (1980b). *Presbyteries and Profits: Calvinism and the Development of Capitalism in Scotland, 1560–1707*. Clarendon Press.

McKinlay, A., & Mutch, A. (2015). Accountable Creatures: Scottish Presbyterianism, Accountability and Managerial Capitalism. *Business History, 57*(2), 241–256.

Mepham, M. J. (1988). The Scottish Enlightenment and the Development of Accounting. *The Accounting Historians Journal, 15*(2), 151–176.

Milevsky, M. A. (2023). Adam Smith's Reversionary Annuity: Money's Worth, Default Options and Auto Enrollment. *Financial History Review, 30*(2), 162–197.

Morren, N. (1838). *Annals of the General Assembly of the Church of Scotland, from the Final Secession in 1739, to the Origin of the Relief in 1752: With an Appendix of Biographical Sketches, Illustrative Documents, and Notes*. J. Johnstone.

Mutch, A. (2017). The Business of Religion: Lending and the Church of Scotland in the Eighteenth Century. *Journal of Scottish Historical Studies, 37*(2), 136–154.

Sefton, H. R. (1962). *The Early Development of Moderation in the Church of Scotland*. University of Glasgow.

Sher, R., & Murdoch, A. (1983). Patronage and Party in the Church of Scotland, 1750–1800. *Church, Politics and Society* (pp. 267–298).

Sher, R. B. (2015). *Church and University in the Scottish Enlightenment* (2nd ed.). Edinburgh University Press.

Sibbett, T. (1993). Book Review: *The Scottish Ministers Widows Fund 1743–1993* by A. Ian Dunlop. *Journal of the Institute of Actuaries, 120*(2), 387–388.

Sibbett, T. (2004). Widows Fund. In Bjørn Sundt & Jeff L. Teugels (Eds.), *Encyclopedia of Actuarial Science*. John Wiley & Sons.

Smout, T. C. (1972). *History of Scottish People, 1560–1830*. Fontana.

Tawney, R. H. (2021 [1926]). *Religion and the Rise of Capitalism*. Angelico Press.

Weber, M. (2002 [1905]). *The Protestant Ethic and the 'Spirit' of Capitalism*. Penguin Books.

CHAPTER 6

Alexander Webster and the Archives

Alexander Webster: Administrator Extraordinaire

The acknowledged father of history, Herodotus of Halicarnassus, writing in the fifth century BC, begins his classic by providing his motives: "So that human achievements may not be forgotten in time, and great and marvellous deeds…may not be without glory." To historians, Herodotus is known for placing human agency at the centre of *The Histories* because he viewed history as the direct result of human actions, decisions, and interactions. People were the primary actors in history, not Homer's gods or divine interventions as in the *Iliad* or *Odyssey*, all of which was a novel idea at the time.

With that introduction in mind, if there is a persona of King Xerxes—the main antagonist of the Greeks—to be found within the eighteenth-century battle to create the (first) Ministers Widow's Fund, it would have to be the Reverend Alexander Webster. This chapter will highlight his behind-the-scenes effort and more generally the operational mechanics of managing the scheme. This is consistent with the thesis of this book that the success of longevity risk pooling and annuities goes far beyond actuarial science. Coincidentally, and speaking of the Greek and Persian wars

M. A. Milevsky, *The Religious Roots of Longevity Risk Sharing*,
https://doi.org/10.1007/978-3-031-62403-2_6

of 490 and 480 BC, the Reverend Alexander Webster—our protagonist—was rather fond of the story of Xerxes and his (apparent) wife, Esther.[1] We know this from the extant transcripts of his (famous) Sunday sermons at Tolbooth Kirk in Edinburgh, where he loved quoting from the Book of Esther. More importantly, Alexander Webster was the administrative force behind the annuity scheme and is why any human-focused story must begin with him.

Alexander Webster was more than just a financial mover and shaker. On spiritual matters, he was a key figure in a period and religious movement known as the Cambuslang Revival, which took place in a village near Glasgow, where a prominent Methodist preacher from England drew thousands of listeners from across Scotland. In particular, Alexander Webster also played a significant role in chronicling these events, and he meticulously recorded the personal stories and transformative experiences of attendees.

It's unfortunate that Webster's glory—be it spiritual or financial—has been overshadowed due to his reputation as a prodigious drinker and perhaps even an alcoholic. His nickname was Dr. Bonum Magnum, Latin for "great good" and, coincidentally, the name given to the giant bottle of champagne. The famed Scottish biographer and diarist James Boswell (b. 1740, d. 1795) wrote that playwright Samuel Johnson (b. 1709, d. 1784) had met Webster for dinner and found him entertaining and agreeable, although "not learned." The few authors who give Webster (some) credit often summarize him as a drunken, garrulous, hot-headed caricature who played second fiddle to the more serious founders of the scheme.[2]

Yet anyone who takes the time to review the handwritten minutes of the trustee's meetings preserved among the National Records of Scotland will quickly be disabused of the notion that Webster's attendance was solely for entertainment. No, he was certainly not a mathematician of the calibre of Colin Maclaurin—the eminent student of Isaac Newton—and he wasn't a demographer or statistician of the calibre of Robert Wallace, the renowned debater of Robert Malthus. If you recall from earlier in the book, Maclaurin and Wallace are the better-known names among modern

[1] The historicity of the Old Testament book of Esther, and in particular the view that King Ahasuerus is Xerxes, has been called into question by biblical scholars. The above note reference simply highlights the fact that Reverend Webster's sermons delivered during the 1740s, all made reference to the story of Esther.

[2] See, for example, Perman (2022, Chapter 6).

aficionados of the scheme. But Alexander Webster was "the guy" who got things done. Every committee has (or should have) one of those people. Yes, the scientists get Herodotus's glory, and their ideas garner Nobel prizes, but a Protestant work ethic is required to put theory into practice. In fact, beyond history, that is the crucial point of this chapter. The success of a retirement plan doesn't hinge on an actuarial formula or algorithm. Instead, mathematics is a necessary but not a sufficient ingredient for sustainability. I might even proclaim that scientists should "invent" and mathematicians should "prove," but they should both get out of the way to make room for the professional administrators—even if the latter are known to enjoy their wine. So, let's begin and delve into our convivial reverend and his contributions to the scheme.

Managing a Meeting of the Fund Trustees

The first meeting of the trustees—most of whom were from Edinburgh, but also included representatives from beyond—included notables such as the historian William Robertson, theologian Hugh Blair, principal William Wishart, and of course Alexander Webster. That very first meeting took place in Edinburgh, on the second Tuesday (the 10th) of April 1744 (old style), between the hours of 10 a.m. and 12 noon. The meeting itself began and ended with a prayer, although God and religion were not to be found anywhere in the discussion between those two points. One knows these facts for certain thanks to the handwritten meeting minutes that have been preserved for nearly 280 years—together with every subsequent meeting of the annuity fund trustees—in the archives of the National Records of Scotland. This chapter will focus on many of the insights that can be gleaned from those minutes.[3]

The first order of business, after electing a *praeses* or chair for the meeting and a clerk to take notes, was to review and process the written notifications that ministers and professors had sent in, in which they actively selected the rates at which they wanted to contribute. And, consistent with the statistics that I shared earlier, approximately 30 of the 35 names mentioned, all selected the highest contribution rate. According to the minutes, three people selected the lower third class and one, Mr. John Glen, sent in a declinature. In other words, he didn't want to be

[3] Recorded from here on as CH9/1/1, which was scanned and digitized by staff at the National Records of Scotland (NRS), for the author, and transcribed by Emma Bromley.

"subject to any of the yearly rates" and—although he had six months to let the trustees know—he made sure to forward a signed resolution to that effect at the first opportunity possible, and quite clearly opted out of the plan.[4] All of this procedure was rather routine and tedious, and even the trustees themselves noted that it was "unnecessary that each of these writings should be recorded at full length."[5]

But the monotony of individual selections eventually made way to some controversy when the matter of eligibility was raised and debated: *Who exactly was entitled to join and contribute to the annuity fund?* While the Act of Parliament used straightforward designations and dates, the question of who exactly was a "minister of the church" or who was defined as a "head, principal or master of the Universities" was left up to the interpretation of the trustees. For example, was an ordained assistant who wasn't married, but who was a designated successor to the present (old, frail) incumbent entitled to join the annuity fund? One such person, Mr. Robert Pollack, sent in precisely that inquiry. Another related inquiry was regarding ministers that had demitted or gave up their post (voluntarily) just prior to the date of 25 May 1744. Where they entitled to participate in the annuity fund if they were "retired" from the Church of Scotland. There were no mediate answers or decisions, and like committees in the twenty-first century, a subcommittee of the trustees was created to deal with these matters. The subcommittee consisted of three trustees, one of whom was, yes, Alexander Webster.[6] His name appears repeatedly, and most frequently, whenever someone volunteered or was volunteered; the recurrent *mover and shaker*.

As the plan notices and declinatures came into the hands of the trustees of the annuity fund, they had to be stored somewhere, together with the financial documents pertaining to the collector for the fund, more of

[4] NRS, https://catalogue.nrscotland.gov.uk/nrsonlinecatalogue/details.aspx?reference=CH9%2f1%2f1&st=1&tc=y&tl=n&tn=n&tp=n&k=&ko=p&r=CH9%2f1%2f1&ro=c&df=&dt=&di=y, CH9/1/1, p. 8.

[5] NRS, https://catalogue.nrscotland.gov.uk/nrsonlinecatalogue/details.aspx?reference=CH9%2f1%2f1&st=1&tc=y&tl=n&tn=n&tp=n&k=&ko=p&r=CH9%2f1%2f1&ro=c&df=&dt=&di=y, CH9/1/1, p. 3.

[6] NRS, https://catalogue.nrscotland.gov.uk/nrsonlinecatalogue/details.aspx?reference=CH9%2f1%2f1&st=1&tc=y&tl=n&tn=n&tp=n&k=&ko=p&r=CH9%2f1%2f1&ro=c&df=&dt=&di=y, CH9/1/1, p. 9.

which will be discussed later. The trustees felt it was necessary to commission the building of a chest where the papers could be "lodged for safety." And, the person who was tasked to find and employ "a fit tradesman to make such repositories" was (yes) Alexander Webster.[7]

With that general background in place, the next few pages offer some additional insights into the level of detail in managing the fund. In the minutes of 74 meetings that I reviewed, Alexander Webster missed only two meetings, during which nothing much was achieved as evidenced by the meagre and short one-page description of those meetings. Before I delve into the details, Table 6.1 lists the names of trustees who appeared most frequently in those 74 meetings, which covered the first decade of trustee minutes stored at the National Records Office (NRS) in Edinburgh.[8]

While Table 6.1 doesn't list every name mentioned in the over-a-decade of trustee meetings, it does mention the names of those who appeared most frequently and, more specifically, the top 20 names. In the top four spots we have Alexander Webster and Robert Wallace (naturally), as well as George Lindsay and George Wishart, who in 1746 was principal clerk of the assembly.[9] There are at least another 20 names over 25 additional meetings, but those names don't appear as frequently. So, for example, the philosopher Thomas Reid, who participated in the scheme, appears only twice. Likewise, Colin Maclaurin, the brilliant and famous mathematician from the University of Edinburgh who "priced" the annuities and determined suitable contribution rates, also only appears twice. The reasons for the dearth of appearances by some of these (eighteenth-century celebrities) will be explained later, but one factor is simply a matter of survival, that is being part of the scheme from 1745 to 1755. Another (celebrity) example of someone who didn't make the cut is

[7] NRS, https://catalogue.nrscotland.gov.uk/nrsonlinecatalogue/details.aspx?reference=CH9%2f1%2f1&st=1&tc=y&tl=n&tn=n&tp=n&k=&ko=p&r=CH9%2f1%2f1&ro=c&df=&dt=&di=y, CH9/1/1, p. 10.

[8] NRS, https://catalogue.nrscotland.gov.uk/nrsonlinecatalogue/details.aspx?reference=CH9%2f1%2f1&st=1&tc=y&tl=n&tn=n&tp=n&k=&ko=p&r=CH9%2f1%2f1&ro=c&df=&dt=&di=y, CH9/1/1, reviewed and counted by author.

[9] Robert Wallace and George Wishart were the two commissioners sent to London in late 1743, to help arrange the formal Act which would make participation in the scheme compulsory. And, since they were still in London, they missed the first meeting of the trustees on 10 April 1744.

Table 6.1 Work of trustees: Appearances, attendance, and activity

Rank	*Last Name*	*First Name*	*Number*	*Chair*
1	Webster	Alexander	162	5
2	Wallace	Robert	111	3
3	Lindsay	George	105	4
4	Wishart	George	92	1
5	Pollock	Robert	81	0
6	Stewart*	James	82	0
7	Hamilton	Robert	78	3
8	Stevenson	James	87	3
9	Logan	George	80	2
10	Blair	Hugh	82	1
11	Walker	James	70	4
12	Wishart	William	69	1
13	Kinloch	Robert	75	2
14	Aitken	William	74	1
15	Gusthart	William	74	1
16	Erskine	John	68	0
17	Pitcairn	Thomas	55	1
18	Fordice	George	53	5
19	Mackie	Charles	57	1
20	Cuming	Patrick	53	2

Source Annual Reports, Meeting Minutes. 1744–1745 NRS CH9/1/1, CH9/16. *Collector of Scheme

William Robertson (b. 1721, d. 1793),[10] the eighteenth-century historian, moderator of the Church of Scotland, and principal of the University of Edinburgh. His name only appeared 35 times in the minutes, and yet he was one of the most influential members of the Scottish Enlightenment.[11] He would have been in his early twenties when the fund was set up, and only recently admitted to the ministry and ordained, so it's unsurprising that during the first decade he was never elected chair (*praeses*).

[10] He would have been in his early 20s when the fund was set up, only just admitted to the ministry and ordained.

[11] See also the book by Thomas Ahnert (2015), on the Moral Culture of the Scottish Enlightenment, the cover of which has the painted image (by Sir Henry Raeburn) of Principal William Robertson.

Either way, one must be careful not to assign too much weight—or relative importance in the scheme hierarchy—to name frequencies. Nevertheless, Alexander Webster had an average of over two appearances per meeting, implying that he was more than just present and answering yes to a trustee roll call; rather, he was the most frequent participant and the most likely to engage with the matters being discussed. He was also most likely (tied with George Fordice) to have been selected as the leader or the "chair" for the trustee meeting a total of five times. And, as anyone who has ever attended a university faculty meeting can attest, this implies that he was *doing more* than just talking. Rather, he might have been selected as the one most likely to keep others from doing too much of that.

We Seek to Avoid "Greatly Embarrassing the Trustees"

The title of this subsection is direct quote from one of the meeting's minutes and captures the ethos that drove the scheme's management.[12] Over the next few pages, you will see how the successful management of the annuity fund involved much more than mathematics. To start, here is an excerpt dealing with trustees and committee members who were absent from the meetings they were supposed to attend. Notice the repercussions.

> The absence of Mr Thomas Turnbull from this meeting is excused for reasons offered. Mr George Logan was absent without sending any excuse, wherefore the meeting find him liable in twenty shillings unless at next meeting he show sufficient cause for his absence.[13]

To place that monetary fine or penalty in perspective, twenty shillings (or a pound) was a quarter of the average annual contribution level to the fund. In fact, every meeting opened with a roll call and a review of the excuses for those who missed prior meetings. These were meetings

[12] NRS, https://catalogue.nrscotland.gov.uk/nrsonlinecatalogue/details.aspx?reference=CH9%2f1%2f1&st=1&tc=y&tl=n&tn=n&tp=n&k=&ko=p&r=CH9%2f1%2f1&ro=c&df=&dt=&di=y, CH9/1/1, NRS, p. 172, meeting 14, dated: 14 May 1745.

[13] NRS, https://catalogue.nrscotland.gov.uk/nrsonlinecatalogue/details.aspx?reference=CH9%2f1%2f1&st=1&tc=y&tl=n&tn=n&tp=n&k=&ko=p&r=CH9%2f1%2f1&ro=c&df=&dt=&di=y, CH9/1/1, NRS, p. 23, meeting 3, dated: 21 May 1744.

that involved ministers and professors travelling from far away, and the excuses that were not accepted by committee members for not attending in the eighteenth century might have some lessons for us in the twenty-first century. Here is one particular excuse from Reverend John Sutherlan that was not accepted as a reason for him not standing in as a trustee:

> There was produced and read a letter dated the fourteenth day of February last by the Reverend John Sutherland, Minister at Tain in the Presbytery of Tain, addressed to the Praeses [*sic*] of the trustees of the widows scheme, requesting that the said trustees would please either accept of that letter as his declinature of the office of trustee or if that could not be agreed to that, they would sustain the distance of one hundred and eighteen miles from Edinburgh as a constant standing excuse for his not attendance on their meetings. The Trustees...find that they cannot sustain the foresaid letter from Mr Sutherland as a declinature of the said office of Trustee and order Mr Sutherland to be enrolled as a Trustee.[14]

Now, the penalty that the trustees imposed (on themselves) for missing meetings paled in magnitude, when compared with penalties imposed on Moderators of the Presbytery who were delinquent in their own responsibilities. Notice the magnitude of this fine, imposed for sloppy note-keeping, and in particular the absence of proper ages for the children who were entitled to payments from the fund:

> There being nothing sent hither from the Presbytery of Abernethy other than a letter intimating the death of Mr. Francis Grant, Minister of Abernethy, and a Presbyterial Certificate of the names of his children without mentioning their ages or other circumstances, the Trustees find the same not satisfying, it being impracticable to make up the Books without the lists of particulars duly attested and therefore as this meeting cannot admit Mr. Grant's children upon the list of annuitants in order to draw the provision to which they would have been entitled at Whitsunday next, had the said lists been duly sent up, the Trustees do in terms of the Act of Parliament find the Moderator of the Presbytery liable in ten pounds sterling and do direct the Collector to levy and apply the same as directed by the statute and that in regard the said lists have not been sent hither, notwithstanding it appears by a Declaration dated the [blank] day of May last under the

[14] NRS, https://catalogue.nrscotland.gov.uk/nrsonlinecatalogue/details.aspx?reference=CH9%2f1%2f1&st=1&tc=y&tl=n&tn=n&tp=n&k=&ko=p&r=CH9%2f1%2f1&ro=c&df=&dt=&di=y, CH9/1/1, NRS, p. 482, meeting 62, dated: 22 May 1753.

> hands of Mr Patrick Grant, Minister at Duthill, that the printed specimen of the lists requisite came to their hands.[15]

Next, let's move on to the careful and precise manner in which financial and demographic information about participants, their contributions, and their beneficiaries were supposed to be collected and stored.

> Mr Webster pursuant to appointment of last meeting produced a specimen of lists to be sent to presbyteries and universities made out under proper columns containing the abovementioned and other facts necessary for the Trustees to have account of for their more easily carrying the said Act of Parliament into execution, which lists being read over with notes thereto subjoined directing the manner of filling up the facts requisite to be known under each particular column.[16]

There reference to "proper columns" and "specimen of lists" comes up repeatedly is discussions of the trustees during the first few years and was eventually brought to the attention of the General Assembly when many of the presbyteries were delinquent in following these instructions. Here is one example of the frustration around properly compiling these lists and the efforts to which the managers and trustees went to inform the presbyteries of their obligations:

> Lists of particular facts necessary to be known to the Trustees were sent off to the several presbyteries and universities about the beginning of last month in common course of post and by Carriers to such places where there are no post offices. Ordered that public advertisement be given in the Edinburgh Newspapers three several days notifying to all concerned that the said lists were accordingly sent off that none pretend ignorance and requiring presbyteries and universities to send due and timeous returns under the penalties contained in the Act of Parliament.[17]

[15] NRS, https://catalogue.nrscotland.gov.uk/nrsonlinecatalogue/details.aspx?reference=CH9%2f1%2f1&st=1&tc=y&tl=n&tn=n&tp=n&k=&ko=p&r=CH9%2f1%2f1&ro=c&df=&dt=&di=y, CH9/1/1, NRS, p. 240, meeting 28, dated: 3 March 1747.

[16] NRS, https://catalogue.nrscotland.gov.uk/nrsonlinecatalogue/details.aspx?reference=CH9%2f1%2f1&st=1&tc=y&tl=n&tn=n&tp=n&k=&ko=p&r=CH9%2f1%2f1&ro=c&df=&dt=&di=y, CH9/1/1, NRS, p. 38, meeting 4, dated: 27 June 1744.

[17] NRS, https://catalogue.nrscotland.gov.uk/nrsonlinecatalogue/details.aspx?reference=CH9%2f1%2f1&st=1&tc=y&tl=n&tn=n&tp=n&k=&ko=p&r=CH9%2f1%2f1&ro=c&df=&dt=&di=y, CH9/1/1, NRS, pp. 52–53, meeting 5, dated: 3 August 1744.

The following echoes the same concerns around the exact structure and composition of the lists:

> As to require these special directions: however, the Trustees must be excused to insist upon an uniform method by Presbyteries in returning their annual lists, without which it is impossible to manage the affairs committed to them, with justice to the Fund or honour to themselves; and I persuade myself you will be of the same mind when I have told you that by reason of the imperfect, irregular lists transmitted to them, their Committee, after several months labour and writing hundreds of letters, have not yet been able to attain the full knowledge of absolute and necessary facts, and cannot finish the plans in their books. You know that the General Assembly, the Society for propagating Christian Knowledge, and indeed all regular Societies whatsoever, lay down certain forms to be observed in attesting facts; and doubtless it is much more necessary in the present than in most cases because the plan is so complex and consists of such a variety of columns - one thing I am likewise sure of, it would not give Presbyteries one half of the trouble to follow the words, order and columns pointed out to them, which some of them now take in writing two or three sheets of paper in every one of their reports - and I am appointed upon the whole to acquaint every Presbytery that if their reports for the future are not made exactly according to the necessary directions given, or which may be afterwards found necessary, the Trustees will be under the necessity of taking some method to ease themselves of the burden and danger of their present Trust or be allowed, without any reflections upon them, to use the powers the Law has put in their hands for obliging Presbyteries to transmit the necessary lists in an uniform, exact manner.[18]

As mentioned above, the question of where exactly to store all of these critical documents and lists once they were compiled came up. Once again Alexander Webster took charge of the matter:

> Mr. Webster reported that, in obedience to the order made the tenth of April last, a strong chest with drawers there for holding the bonds and other papers of value are prepared, together with a separate desk in which are certain conveniences for holding the Trustees' Books, paper etc. above

[18] NRS, https://catalogue.nrscotland.gov.uk/nrsonlinecatalogue/details.aspx?reference=CH9%2f1%2f1&st=1&tc=y&tl=n&tn=n&tp=n&k=&ko=p&r=CH9%2f1%2f1&ro=c&df=&dt=&di=y, CH9/1/1, NRS, pp. 265–266, meeting 29, dated: 12 May 1747.

which is a press consisting of eighty-eight divisions for holding the notifications, Declinatures and Lists to be sent up from the several presbyteries and universities distinctly by themselves; the said Repositories being brought to the Society Hall and now inspected by the Trustees they are satisfied therewith and remit to Mr. Webster to agree with Mr. Wright as to the price of the said piece of work.[19]

Notice the 88 divisions within the "strong chest" for holding the notification and lists! Once the trustees reviewed and inspected said chest, permission was granted to Webster to negotiate over the price. As the meetings of the trustees evolved into regular gatherings at fixed times, it was determined that a special location should be built especially for that purpose, partially because it was viewed as safer.

The Committee, formerly appointed by the Trustees for considering of several particulars referred to them, brought in a memorial bearing the inconveniences the Trustees are under by meeting in the Old Kirk Isle or the Society's Hall, and the necessity of having a house built distant from any other building with a Hall for accommodating the Trustees at their meetings, in which the Offices of Collector and Clerk may be under one roof and other conveniences had for holding the papers and repositories of the Trustees, whereby the same may not be in hazard of being lost by fire etc. and proposing that the foresaid sums granted by the Earl of Stair and Mr Andrew Mitchel may be applied for that purpose which, together with other contributions that may be got, and the sum of seven pounds, ten shillings by the Collector and two pounds, ten shillings by the Clerk, which they are willing to pay yearly for their respective Offices, will be sufficient to undertake such a work. The Trustees, having heard the said memorial, unanimously resolved that the said sums legated by the Earl of Stair and Mr Andrew Mitchel be applied for, defraying the expense of said building, and remit to their Committee to look out for a proper area and prepare a plan and estimate of the building to be laid before another meeting.[20]

[19] NRS, https://catalogue.nrscotland.gov.uk/nrsonlinecatalogue/details.aspx?reference=CH9%2f1%2f1&st=1&tc=y&tl=n&tn=n&tp=n&k=&ko=p&r=CH9%2f1%2f1&ro=c&df=&dt=&di=y, CH9/1/1, NRS, pp. 83–84, meeting 7, dated: 27 September 1744.

[20] NRS, https://catalogue.nrscotland.gov.uk/nrsonlinecatalogue/details.aspx?reference=CH9%2f1%2f1&st=1&tc=y&tl=n&tn=n&tp=n&k=&ko=p&r=CH9%2f1%2f1&ro=c&df=&dt=&di=y, CH9/1/1, NRS, pp. 105–106, meeting 9, dated: 2 January 1745.

It's interesting to note that the new building in which trustees met to discuss the management of a fund to protect against longevity risk, was carefully insured against another risk, fire:

> The Trustees, considering that it will be of advantage that their hall and other apartments thereto belonging, lately fitted up, be insured against losses by fire, do authorise their Committee as they shall after enquiry judge reasonable to insure the same, either at the Sun Fire Office, London, or at the Edinburgh Insurance Office.[21]

And, speaking of oversight and the micro-management of affairs by the trustees, the following excerpt speaks to the matter of strictness around time limits and deadlines. Recall that participants had to note their choices before the deadline or they would be defaulted into the second class. Well, here is an example of someone who sent their notice in late, but clearly indicated that he had changed his mind and wanted to be in the third class rather than the second class:

> Read a notification dated the twenty-fourth of July last by Mr. Edward Steedman, Minister at Haddington, of his choice of the rate of five pounds, five shillings sterling but it being only yesterday the twenty-sixth instant being a day without the time directed by the Act of Parliament delivered in at the Collector's Office; the Trustees reject the same and ordered that Mr Steedman be placed in the class of three pounds, eighteen shillings yearly. Notifications by certain persons who had formerly notified by writings under their hands their choice of rates therein mentioned and having since changed the classes to which they at first acceded insist upon their right to make such alteration of their choice, the Trustees, after reasoning upon this question, did resolve that a person, once notifying his choice of a rate by a writing under his hands duly executed and entered in the Collector's Books as the law directs, cannot by an after deed alter or revoke the former one and that a person, once declining to be subject to any of the rates or other payments in the Act of Parliament mentioned, cannot by

[21] NRS, https://catalogue.nrscotland.gov.uk/nrsonlinecatalogue/details.aspx?reference=CH9%2f1%2f1&st=1&tc=y&tl=n&tn=n&tp=n&k=&ko=p&r=CH9%2f1%2f1&ro=c&df=&dt=&di=y, CH9/1/1, NRS, p. 306, meeting 34, dated: 5 April 1748.

an after deed or writing accede or be entitled to be a contributor to the Fund or enjoy any benefit arising therefrom, and accordingly.[22]

Another example of the diligence and care with which every single participant in the fund was carefully vetted to determine whether they were eligible to participate is found in this excerpt from a debate of the trustees on whether professors who didn't get paid (for whatever reason) were allowed in:

> Professors having no salaries after reasoning upon the case of professors or masters in Colleges who have no settled salaries, the Trustees remit to the Committee to advise with Lawyers how far they should be received as Contributors to this Fund and report.[23]

Not to be forgotten in all of this, is that these meetings took place in the dead of winter, in Edinburgh, which could get very cold and damp.

> Mr Webster was unanimously chosen praeses [sic] and began with prayer, after which the Trustees agreed to retire from the Old Kirk Isle to the Hall of the Society for propagating Christian Knowledge by reason of the coldness of the season.[24]

Beyond the Scottish weather, the Scottish strictness with which the fund was managed, and the micro-management of the participants, their choices, and their contributions is especially evident from cases in which the trustees suspected fraud or malfeasance. Here is just one example of a situation in which this occurred:

> Mr James Weir: It being observed by the Committee that, upon inspection, it appears the notification of the now deceased Mr James Weir, Minister at Orphire in Orknay, though mentioned to be written with his own hand,

[22] NRS, https://catalogue.nrscotland.gov.uk/nrsonlinecatalogue/details.aspx?reference=CH9%2f1%2f1&st=1&tc=y&tl=n&tn=n&tp=n&k=&ko=p&r=CH9%2f1%2f1&ro=c&df=&dt=&di=y, CH9/1/1, NRS, pp. 86–87, meeting 7, dated: 27 September 1744.

[23] NRS, https://catalogue.nrscotland.gov.uk/nrsonlinecatalogue/details.aspx?reference=CH9%2f1%2f1&st=1&tc=y&tl=n&tn=n&tp=n&k=&ko=p&r=CH9%2f1%2f1&ro=c&df=&dt=&di=y, CH9/1/1, NRS, p. 95, meeting 8, dated: 13 November 1744.

[24] NRS, https://catalogue.nrscotland.gov.uk/nrsonlinecatalogue/details.aspx?reference=CH9%2f1%2f1&st=1&tc=y&tl=n&tn=n&tp=n&k=&ko=p&r=CH9%2f1%2f1&ro=c&df=&dt=&di=y, CH9/1/1, NRS, p. 128, meeting 11, dated: 12 February 1745.

> yet was actually written by another. The Trustees discharge the Collector to receive any payment from his heirs or others for them 'til further orders.[25]

Recall that the biggest financial risk to the fund was the event in which contributors paid very little (i.e., died quickly) and then their beneficiaries would be entitled to annuities and payments that hadn't been fully paid for. So, ensuring that people who had elected to participate were actually alive at the time the election was made was critical. And, while the above excerpt is polite enough not to accuse the beneficiaries of any shenanigans, the sentence "…was actually written by another" says it all. Another aspect of the oversight and governance of the fund by the trustees is the transparency and openness with which matters were dealt with. Notice in the following excerpt that once the lists were compiled they were made available to the public:

> Trustees being by Act of Parliament appointed on the 2nd Tuesday in March yearly to make up proper lists of all the widows and children entitled to receive annuities or provisions out of the Funds of the preceding year and a list being accordingly made out and laid upon the table, the Trustees, after inspecting the same, ordered that said list remain in the hands of the Clerk and be open to the inspection of all parties interested to the end they may give in their objections in writing to the Clerk if they have any, which he is ordered to lay before the Trustees the 2nd Tuesday of May next.[26]

It's quite clear that these proceedings were not opaque back-office dealings, and anyone who objected to anything on the lists of beneficiaries could complain to the clerk. It's worth recalling (once again) that this took place in the eighteenth century, well before the advent of insurance regulators, ombudsmen, and anonymous whistle-blowers. Notice in the following excerpt, the instructions assuming there were no objections, on the duplicate copies and procedure for disbursement of funds:

[25] NRS, https://catalogue.nrscotland.gov.uk/nrsonlinecatalogue/details.aspx?reference=CH9%2f1%2f1&st=1&tc=y&tl=n&tn=n&tp=n&k=&ko=p&r=CH9%2f1%2f1&ro=c&df=&dt=&di=y, CH9/1/1, NRS, p. 110, meeting 9, dated: 2 January 1745.

[26] NRS, https://catalogue.nrscotland.gov.uk/nrsonlinecatalogue/details.aspx?reference=CH9%2f1%2f1&st=1&tc=y&tl=n&tn=n&tp=n&k=&ko=p&r=CH9%2f1%2f1&ro=c&df=&dt=&di=y, CH9/1/1, NRS, p. 142, meeting 12, dated: 12 March 1745.

> The above list of the disposal of the widows Fund, having remained in the Clerk's hand until this day and no objections being given in, the Trustees do order that their Praeses [*sic*] sign two fair copies thereof, one of which to be forthwith given to the General Collector for payment accordingly in favor of each widow and child interested and the other to lie in the Clerk's hands.[27]

Note that this transparency and openness with regard to who was receiving payments also applied to those (presbyteries) who were not making their own contributions, and in time. Notice the following admonishment:

> Arrears of vacancies it appearing upon inspecting the Collector and Clerk's books that a great many Presbyteries are deficient in paying the taxes due at Whitsunday last out of the vacancies in their respective Bounds, which should have been at furthest paid on the fifteenth of November last, ordered that public advertisement be given in the newspapers requiring them to make speedy payment, otherwise there will be a necessity of ordering a prosecution against them.[28]

As time went on, these admonishments and complaints grew louder. Clearly, some presbyteries were (so) tardy and disorganized with the information and the list they had to submit that this matter was getting personal. The trustees' reputation was at stake and the long-term viability of the fund itself was called into question:

> Representation to the General Assembly about keeping a separate register The Trustees, considering the disadvantages arising from the delay of presbyteries in sending up their lists and not keeping separate registers in an uniform way, ordered a representation to be laid before the Assembly this day by their Praeses [*sic*], Masters Mathison, Dickson, Wallace and McKenzie, craving that the Assembly would renew the Act fourth14 of last Assembly, obliging each presbytery to keep a separate register for the purposes therein contained and also to record therein the other facts, the

[27] NRS, https://catalogue.nrscotland.gov.uk/nrsonlinecatalogue/details.aspx?reference=CH9%2f1%2f1&st=1&tc=y&tl=n&tn=n&tp=n&k=&ko=p&r=CH9%2f1%2f1&ro=c&df=&dt=&di=y, CH9/1/1, NRS, p. 206, meeting 22, dated: 13 May 1746.

[28] NRS, https://catalogue.nrscotland.gov.uk/nrsonlinecatalogue/details.aspx?reference=CH9%2f1%2f1&st=1&tc=y&tl=n&tn=n&tp=n&k=&ko=p&r=CH9%2f1%2f1&ro=c&df=&dt=&di=y, CH9/1/1, NRS, p. 144, meeting 12, dated: 12 March 1745.

As yet another example of the trustees' strict adherence to the letter of the Act, the following provided an interesting specimen because it collides with that "other" event that took place in Scotland during the period 1744 and 1745. Read on:

> A notification dated the seventeenth April by Mr John McVicar, Minister of Islay, of his choice of the rate of six pounds, eleven shillings and three pence being read, marked as produced the twenty-sixth of September last. There was given in a memorial for the said Mr McVicar bearing his residing in an island from which he had in April last sent the said deed to a friend on the mainland to be by him forwarded timeously to the Collector but that he had forgot it 'til September when an Express was dispatched for Edinburgh with it, who was interrupted and detained for some days by the rebels near Glasgow, by which means he could not reach Edinburgh 'til the twenty-sixth of September, the day after the eighteen calendar months mentioned in the Act of Parliament expired. The Trustees, having considered the said notification and memorial, did reject the same in regard there was full time for transmitting hither the notification before the troubles in the country began and ordered that Mr McVicar be stated as neglecting to notify his choice and therefore subjected to the annual rate of three pounds, eighteen shillings and nine pence and other payments consequent thereupon.[32]

The Jacobite Rebellion and Bonnie Prince Charlie himself weren't deemed to be a valid excuse for Mr. McVicar's tardiness. According to the trustees, he should have gotten his notification sent before the deadline. He wanted to contribute at the highest possible rate and be included in the fourth class, but the trustees demoted him to the (default option) second class. As far as they were concerned, he didn't send the paperwork in on time. Period.

Revisions to the Scheme in 1748

For technical reasons that will be better explained and understood in the next chapter, the trustees were forced to tinker with some of the parameters of the scheme, although not in any obvious manner. The four

[32] NRS, https://catalogue.nrscotland.gov.uk/nrsonlinecatalogue/details.aspx?reference=CH9%2f1%2f1&st=1&tc=y&tl=n&tn=n&tp=n&k=&ko=p&r=CH9%2f1%2f1&ro=c&df=&dt=&di=y, CH9/1/1, NRS, p. 181, meeting 17, dated: 28 November 1745.

contribution classes and benefits never changed—nor did the 4% interest rate charged to ministers—but in a round of revisions that took effect in 1748, some tweaks were implemented. In particular, *who* exactly qualified for annuity benefits as well as *how* long someone had to contribute before his widow was entitled to benefits, were tightened. Here is a glimpse into some of those changes, in the words of the trustees themselves:

> First, that proportional deductions shall be made from the provision payable to each family of children at Whitsunday one thousand, seven hundred and forty-nine or any term thereafter, and from the annuity of each widow payable for the first time at the said term of Whitsunday, one thousand, seven hundred and forty-nine or any term thereafter, whose father or husband shall not have paid for support of the Fund a sum equal to three years of the annuity corresponding to the annual rate, to which he was subjected - that is to say - there shall be deduced from the annuity due to every such widow and from the provision due to every such family of children such a sum as together with the rates paid or due by her husband or their father (without computing interest thereon) shall make up or be equal to three years of the said annuity—provided always that for the support of such widows in the meantime, such deductions from their annuities shall be made by retaining only one half thereof 'til the said deficiency be made good to the Fund.[33]

Whitsunday (short for White Sunday) is the weekend associated with the Christian holiday of Pentecost, itself a Greek word (*pentekoste*) which means fiftieth, representing 50 days after Easter. The exact date fluctuates from year to year (depending on the solar versus lunar cycle) and is usually in late May or early June. Whitsunday was the day, or the Sunday, when all payments were made to annuitants and/or children from the fund. They were paid twice per year, with the other date being Martinmas, which is St. Martin's day in November. Either way, as for the year 1749, the payment terms were tightened. If the contributor had not made three full payments to the fund prior to his death, the reversionary annuity benefit would be reduced by the unpaid amount until three years of contributions were caught up. This was a minor tweak but ensured that anyone who joined the plan and then promptly died would have to make up some of the "loss" to the fund after his death.

[33] CH9/1/1, NRS, p. 336, meeting 36, dated: 22 September 1748.

Choices Made and Contributions Set

Readers should now have a good sense of the day-to-day management of the scheme, and an understanding of some aspects of the scheme that eventually had to be modified. I'll return to the archives again in the next chapter when I discuss some of the investment and lending aspects of the fund, but in the next few pages I'll examine some of the active contribution choices that initial members selected. Recall that those participants could (effectively) select one of three things. They could (1) opt out entirely, (2) do nothing and get defaulted to the second £15 annuity class, or (3) actively select one of the four classes: the lowest at £10; the default option at £15; the second highest option at £20; or the most lucrative reversionary annuity at £25 for their widow. Now, the goal here isn't to exhaustively enumerate the choices made by every celebrity of the Scottish Enlightenment, and the National Records of Scotland cited at the end of this chapter contain the full list of names and choices for those interested in locating celebrities or perhaps ancestors. Rather, the point of highlighting just a few names in the Trustees' Scheme Books—from approximately 65 pages covering the period from 1745 to 1755—is that they shed light on other social and financial aspects of the participants themselves. It also gives us an opportunity to examine their individual contributions to the scheme. I'll begin with one of the architectural triumvirate, Colin Maclaurin, who might have the claim to being one of the world's first actuarial pension consultants.

Colin Maclaurin (b. 1698, d. 1746) is listed in the scheme books as being a Professor of Mathematics at the University of Edinburgh, although in the documentation his name is spelled as McLaurin. Next to his name is the letter "m" which denotes marriage, and indeed (yes) his wife's name was Anne. As someone who was an incumbent prior to 25 March 1744, that is had the option to opt out of the scheme if he wanted, he actively requested to join the scheme. In fact, it would have been shocking had he opted out, considering the fact that he did the original actuarial work and is likely the individual that recommended and advocated that the University of Edinburgh join a scheme originally intended for ministers of the church. (See the next paragraph for more.) Interestingly, Colin elected to contribute at the third annuity class, that is, he committed to paying £5 5s. for a widow's reversionary annuity of £20. Why not select the highest class? We don't know and reasons for selections weren't offered. But, it is known that he was a man of few means, and that the great Sir Isaac Newton subsidized his salary and wages at the university. So, perhaps these tight financial constraints forced him into the cheaper option. Unbeknownst to him at the time of his choice, his wife whose full name was Anne Stewart would be a very early beneficiary of the scheme, since Colin Maclaurin (sadly) died at the age of 48, in the aftermath of the Jacobite'45 uprising. He was injured trying to defend the city against the earlier mentioned Bonnie Prince Charlie.[34]

Reverend Robert Wallace—a third member of the triumvirate—had asked Maclaurin for an opinion on the soundness of the scheme, hoping to use the famous mathematician's authority to gain approval for the Act.[35] Robert Wallace was well-versed in mathematics, having learned from and studied under Professor James Gregory at the University of Edinburgh. Colin Maclaurin replied (3 June 1743) with minor suggested improvements and concerns, such as warning the "scheme was remarkably advantageous to the old" and recommending a tax on ministers who were already married. He seemed aware of earlier micro schemes, which I'll discuss in greater length in Chapter 8, since he wrote the following in a letter to Robert Wallace:

> A greater improvement can be made of large sums and with less danger from the hazards to which all things are subjected by faithful trustees, than of small annual sums by single ministers. As it is a certain rule that no single man unless he be extremely rich ought to deal in insurance, but rich men or companies of men only, because loss to a poor man is more sensible than an equal gain.[36]

Getting to the heart of longevity risk pooling, notice Colin Maclaurin's rather subtle statement about "rich men or companies of men" versus the "single man," being suitable for dealing with insurance. Overall, he clearly approved of the scheme and the following quote from his letter to Wallace appears in a number of places: "I think myself obliged to say that the design is so good that minute objections against the absolute

[34] For Colin Maclaurin, see entry by Erik Lars Sageng in the *Oxford Dictionary of National Biography*. https://doi.org/10.1093/ref:odnb/17643.

[35] Dow (1975), Deucher (1896), as well as Dunlop (1992) all mention his endorsement as being important and one of the reasons the Act was passed with little objection.

[36] CH9/17/16, "Letter from Colin Maclaurin to the Reverend Robert Wallace, 23 May 1743".

perfection of the scheme seem to be improper,"[37] and, as noted earlier, he petitioned to have his colleagues join.

When discussing Colin Maclaurin's contributions to science more generally and to pension and annuity matters more narrowly, the historian Olivier Bruneau writes, "It was thanks to Maclaurin's mathematical treatment that this project was a success." Likewise, Judith Grabiner argues, "The success of the fund stemmed from the way Maclaurin joined mathematical modeling and empirical data...to solve an economic problem for the church and the universities in Scotland."[38] Indeed, the mathematics was part of the success, but not all of it. There are other unsung management heroes. As I have attempted to show over this entire chapter, the claim that a simple equation or the multiple 3.81 "solved the problem," while ignoring the immense administrative infrastructure, might overstate the contribution of "scientists" over "management." More on the problems with a 3.81 multiple between benefits and contributions will come in the next chapter.

Francis Hutcheson (b. 1894, d. 1746) was Professor of Moral Philosophy at University of Glasgow, also known as the intellectual inspiration for the Scottish Enlightenment, and in particular his influence on the economist Adam Smith, who taught at Glasgow in the 1750s. The "not to be forgotten" Hutcheson was an incumbent in 1744, and he decided to opt out of the plan entirely. His notice of declinature was received by the clerk in June 1744, which arguably might display a certain lack of "moral sense." Once again, we have no record of why he made this decision to not purchase a (rather cheap) reversionary annuity for his wife, and in hindsight it would have been a rather good investment. Why? Francis Hutcheson died on 8 August 1746, soon after the scheme went into effect, which is just two months after Colin Maclaurin. One reason for Hutcheson's declinature, might be the following. According to his entry in the *Oxford Dictionary for National Biography,* he left his wife a life annuity of £20 backed by lands he owned in Ballyhackmer and Longford. So, she wasn't left destitute, and one could speculate he declined

[37] Ibid.

[38] Grabiner (2002), "Maclaurin and Newton," p. 152.

to participate due to his acrimonious relationship with the Glasgow Presbytery or perhaps his extensive land holdings.[39] In fact, his colleagues at Glasgow, Professor of Mathematics Robert Sim(p)son[40] and Botany Professor Robert Hamilton, both declined to participate in the scheme, as did Philosophy Professor John Stevenson (at the University of Edinburgh). Now let's examine the "most famous" faculty member at the University of Glasgow, a student of Francis Hutcheson. Our next contributor wasn't among the inaugural class (in 1744) but he certainly was a member of the scheme and worthy of note.

Adam Smith (b. 1723, d. 1790), the author of the *Wealth of Nations*, joined the University of Glasgow as a Professor of Logic and Metaphysics in 1751. Oddly, in the Trustees' Scheme Books he is listed (only) as a Professor of Philosophy with a date of admission[41] of 16 January 1751. Smith is then listed as having transferred on 29 April 1752 to become a Professor of Moral Philosophy, which indeed he held from 1752 to 1765. Adam Smith made an active decision—that is, he sent in an official notice—to participate in the scheme at the highest possible contribution rate. This choice is consistent with favourable comments he had made about Alexander Webster in some of his published letters. To be clear, Smith committed to contribute £6.11s. 3d. for the duration of his participation. It's worth noting that William Cullen, whom Nicholas Phillipson[42] called "Smith's main informant on University matters," also selected the highest class. But Smith's competition for the position at Glasgow, George Muirhead, selected the lower third class.[43]

Smith's active choice is rather interesting. At the age of 28 he selected the largest possible annuity for a (yet) non-existent wife. There has been much (frivolous) discussion about the fact that he never married.[44] It's

[39] For Francis Hutcheson, see entry by James Moore in the *Oxford Dictionary of National Biography*. https://doi.org/10.1093/ref:odnb/14273.

[40] Perhaps he held a grudge against Alexander Webster, whose father James prosecuted and persecuted his uncle.

[41] The *Glasgow Courant* noted his start date as 7 January 1751.

[42] Phillipson (2010).

[43] Note the reference to Philosophy (top) and then Moral Philosophy (bottom), but not Logic or Metaphysics.

[44] For Adam Smith, see the entry by Donald Winch in the *Oxford Dictionary of National Biography*. https://doi.org/10.1093/ref:odnb/25767. Also see Phillipson (2010, p. 120).

difficult to reconcile innuendoes with the financial commitment he made by joining the scheme. If indeed he had no intention of ever marrying (in 1751), although he couldn't decline to participate in the scheme—like the "never-to-be-forgotten" Francis Hutcheson—he could have selected the lowest possible class.

Perhaps Smith's annuity choice is just a trivial fact, or the sum itself was trivial relative to his professorial income of £150 to £300 per year,[45] but his active participation and choices have not been noted by any of the multitude of historians or economists who specialize in all things Smith. Now let's examine the choice made by his intellectual nemesis.[46]

Thomas Reid (b. 1710, d. 1796) is a seminal figure in the Scottish Enlightenment and is known as the founder of the Scottish School of Common Sense. His philosophical work significantly influenced the fields of epistemology, the theory of perception, and moral philosophy, and was an intellectual sparring partner of David Hume, the (most famous) Scottish historian and philosopher. Reid's philosophy also influenced the areas of analytic philosophy, logical atomism, and logical pluralism, all of which are well outside the scope of this book (or my abilities). We do know that Thomas joined King's College in Aberdeen in 1751 as a Professor of Philosophy, which was after the act went into effect on 25 March 1744. So, he had no free will in this matter and was forced to participate in the annuity scheme. The epistemological basis for this knowledge is that the 1751 start date for his professorship places him within the 10-year period (1744–1755) covered in the Trustees' Scheme Books, which I can empirically attest to have seen. In fact, he eventually replaced Adam Smith at Glasgow in 1764, which must have been a difficult act to follow. As far as family is concerned, and the beneficiary of his annuity, he married (his cousin Elizabeth) in 1740, which is also carefully noted in the Trustees' Scheme Books. However, and this is the key point, he neglected to (ever) notify the trustees or the clerk of his choice and was therefore—as readers should know by now—"defaulted" into the second class. Reid lived for an additional 45 years after joining the scheme in 1751, so perhaps his absent-mindedness (or lack of choice) worked out in his favour *ex post*. It would have been even more costly for him to have participated in the

[45] Source: Phillipson (2010, p. 182).

[46] Duncan and Baird (1977), "Thomas Reid's Criticisms of Adam Smith's Theory of the Moral Sentiments".

third of fourth class, although that is obviously *ex post* logic. For more on the intellectual life and career of Thomas Reid, see the chapter by the historian Paul Wood, in the 2015 book published by Oxford University Press entitled the *History of Scottish Philosophy*.[47]

William Robertson (b. 1721, d. 1793) was a leading figure of the Scottish Enlightenment. He was a historian, theologian, and principal of the University of Edinburgh, and made significant contributions to historical writing and the intellectual culture of his time. He was also a leader at the University of Edinburgh, which helped to cultivate a unique intellectual environment. According to his entry in *Britannica*, he "is regarded, along with David Hume (https://www.britannica.com/biography/David-Hume) and Edward Gibbon (https://www.britannica.com/biography/Edward-Gibbon), as one of the most important British historians of the eighteenth century." His most notable books were *The History of the Reign of the Emperor Charles V* (1769) and *The History of America* (1777). Robertson was born in Midlothian, educated at Dalkeith Grammar School before attending the University of Edinburgh, and was ordained a minister in the Church of Scotland in 1744, nearly a month before the scheme went into effect.

The historian Richard Sher notes that "in no other European country, Protestant or Catholic, was the Enlightenment so closely tied to, and supported by, the established church and national universities as in Scotland during the age of William Robertson." An annuity scheme allowing only for ministers and professors, which would probably be unheard of in the twenty-first century, attests to that close connection.[48] Now, since he was ordained and joined the ministry just after the scheme went into effect, Robertson did not have the choice to opt out, like Francis Hutcheson did. And in fact, Robertson actively selected the highest annuity class in a notice sent to the trustees in early February 1745. He wasn't married at the time he was ordained, so he contributed the £6.11s. 3d. as a bachelor for over six years, and after he married his cousin Mary on 22 August 1751, he continued to pay at that rate for another 42 years until he died in 1793. Mary outlived him by another 10 years and presumably received the £25 reversionary annuity for the rest of her life. I leave the act of computing the internal rate of return on Robertson's annuity—pay the

[47] Wood (2015).

[48] Sher (2015).

£6.11s. 3d. premium for 48 years and collect the £25 for 10 years—as an exercise for readers, and something I'll return to in Chapter 12.[49]

For the sake of completeness—especially for those who study the Scottish Enlightenment—I'll mention here that William Wishart (secundus), principal of the University of Edinburgh (two prior to Robertson) and who was a moderator of the Church of Scotland in 1745 as the scheme was implemented, selected the highest class. But it appears his wife, Margaret, remarried soon after he died in 1753, which meant that she didn't draw the £25 annuity for long.

Robert Wallace (b. 1697, d. 1771) was a Presbyterian minister and also a notable figure in the Scottish Enlightenment, particularly recognized for his contributions to demography. Born in Kincardine, Wallace was educated in mathematics at the University of Edinburgh. Wallace is best known today for his significant contribution to demography in his 1753 treatise "Dissertation on the Numbers of Mankind in Ancient and Modern Times." In that article he argued against a prevailing theory circulating in the eighteenth century that the world's population was declining. Instead, Wallace suggested that the population could grow under the right conditions of economic and social improvement, anticipating ideas of population dynamics that would later be more fully developed by Thomas Malthus and others. As a member of the Scottish Enlightenment, Wallace was part of a circle that included David Hume and Adam Smith, as well as Robertson and Reid, and he was a co-founder of the Rankenian Club, in which they all met. Despite his contributions to demography and his involvement in the intellectual currents of his time, Wallace remained primarily a religious figure, and has been the subject of a Ph.D. thesis as recently as 1973, written by Norah Smith at the University of Edinburgh. Yet his involvement with the fund was multidimensional and one could argue that he deserves just as much credit as Alexander Webster, even though his name doesn't appear as often in the trustee minutes. Beyond name counts, Wallace is also known for an earlier proposal to create a similar annuity scheme for "Gentlemen" that didn't have land estates. Presumably that proposal (circa 1740) didn't go very far, and he made some changes, consulted Colin Macluarin, and grafted the plan on to the Church of Scotland. Here are some aspects

[49] For William Robertson, see entry by Jeffrey R. Smitten in the *Oxford Dictionary of National Biography*. https://doi.org/10.1093/ref:odnb/28861.

of Wallace's earlier proposal, which yet again speak to the slow, gradual process by which the 1744 scheme came into being.

In a handwritten "Proposal for raising a fund for the widows of Ministers or others," written around 1740 or 1741, and certainly well before the establishment of the 1744 fund for widows and orphans, Wallace wrote the following as part of a much larger document shelf marked CH9/17/10 at the National Records Office:

> Proposals for raising a fund for jointures to the widows of such Gentlemen as have not land estates nor great sums of money but live by their business or yearly income which depends on their lives...It is proposed that a society be formed of the husbands adventurers the contributary to this fund....That every husband pay in yearly to this society what sum he thinks proper during his life & that his widow shall draw after by Death a proportionable sum during her life...its experience alone & a nice calculation that must determine the proportional sum the widow is to have after the husband's death, but a beginning may be made by allowing triple the sum the husband paid in during his life so if a husband pays yearly £10 the widow has £30 if he pay £30 she draws £90 during her life.[50]

There are a few interesting things worth noting, in this earlier proposal, in particular the multiple of 3 between contributions and benefits, versus the final multiple of 3.81 used by the fund. Here is the (crude) actuarial logic that Robert Wallace used to justify the multiple of 3:

> The husband has as good if not a better chance to be the survivor so the half or more than the half of the widows will want no jointure which alone makes the society able to give Double or near to Double of what the husband pays for by the exactness fee calculation the society is not to be allowed to be very rich but to depend chiefly on their yearly incomes & not in the rents of Lands or interests of money. But besides this as the ordinary time of marrying & Consequently of entering into this society (as we shall see afterwards) is about 30 years of age & persons at 30 by the bills of mortality may be supposed to live about 28 years, this made the common time that married persons live together about 28 years, & 58 to be the common time when a woman is a widow. but persons of 58 years of age by the bills of mortality have an equal chance to live only 13 years which is not half the time that married persons live together (that is

[50] CH9/17/10, p. 1. Note that this was transcribed for the author, by Emma Bromley at *Historical Documents Transcription Services*.

> to say the society has only half the number of widows to provide and this half only half the time of their husbands contributing yearly to the society; that is every widow may have the Quadruple of what the husband pays in therefore a triple cannot be too much.[51]

The reasoning was rough, approximate, and completely devoid of any mention of interest rates or the time value of money. But, Robert Wallace was aware of some of the problems that could arise:

> It is proposed therefore that if the husband be not above 10 years older than his wife, he be allowed to enter into this society according to the above proposals. But, if the husband be more than 10 years older than his wife besides the yearly sum he pays, for every year above 10 he must give a premium for his privilege of entering into the society. This premium is to be paid all at once at entering into the society or at the outmost at 3 terms.[52]

How was this premium supposed to be computed? The next few paragraphs go on to discuss, but not conclude, by similar back-of-the-envelope arguments that there should be some additional proportional part. The same proposal makes it very clear that anti-selection would not be tolerated:

> But no man must be allowed to enter upon his death bed or when its visible that he is dying to prevent which it may be constituted that every man must not only come to transact this with the directors but that the bargain shall not be finished till 3 months after the first proposal & the man be personally present. But if he do not enter within a year of his marriage or Constitution of the society besides all other regulations concerning him he must live as [illegible] year after he enters into the society without which his widow shall not be entitled to her jointure.[53]

The proposal then goes on to discuss the minimum size of the pool or fund, which is at least 20, that there should be 30 directors, that they should meet annually, that the society should keep books, and that directors should be rotated, all of which made their way in one form or another

[51] CH9/17/10, p. 2.
[52] CH9/17/10, p. 3.
[53] CH9/17/10, p. 4.

to the 1744 fund for widows and orphans. Indeed, it is quite appropriate that Robert Wallace be viewed as one of the triumvirate of architects and engineers of the scheme. And, quite perceptively, towards the very end of the proposal, he writes: "No man nor his widow shall have it in his power to transfer his or her right to the jointure: to prevent stock jobbing."[54] In other words, the annuity contract was non-transferable. The contributor not the beneficiary was able to sell the annuity after it had been purchased, and the rationale for the lack of a secondary market for this life insurance policy (in modern terms) was that they wanted to avoid the practice of "stock jobbing," which referred to the practice of buying and selling stocks for short-term profit. The practice of "stock jobbing" had very negative connotations, being that this was written in the aftermath of the South Sea Bubble of 1720. For those readers interested in learning more about Robert Wallace, I would suggest the recent (2024) article by Elad Carmel—who laments that Wallace "remains a largely overlooked figure in the literature"—which was published in the *History of European Ideas*.[55] Finally, no list of individual selections would be complete without mentioning the third member of the triumvirate, who, as it turns out, is also our protagonist.

Alexander Webster (b. 1707, d. 1784) was elected moderator of the General Assembly of the Church of Scotland in 1753, conducted the earliest census[56] of Scotland in 1755, and became General Collector of the scheme in 1771. He was also noted as a Popular (and popular) Evangelical preacher,[57] of whom the following was said regarding his Sunday sermons: "It is easier to get into the kingdom of heaven, than a seat in Tolbooth church."[58] As mentioned earlier, he also enjoyed a convivial lifestyle and many sources call him a prodigious drinker who could hold his liquor (or claret) better than any. The diary of Alexander Carlyle, a figure who had run-ins with Webster, nicknamed

[54] CH9/17/10, p. 7.

[55] Carmel (2024)

[56] Riddell (2021), and Anderson (2011) for concerns over the accuracy of Webster's methodology, which is worth noting in the context of the scheme.

[57] For Alexander Webster, see entry by Margaret Stewart in the *Oxford Dictionary of National Biography*. https://doi.org/10.1093/ref:odnb/28939, as well as the eulogy by Anonymous (1784, p. 61).

[58] Mackie (1956), who lavishes praise on Webster and describes how the scheme was copied by Presbyterians in Philadelphia circa 1761.

him "Dr. Bonum Magnum," calling him a "Proteus-like character" who was "extremely pernicious."[59] In the biography of Samuel Johnson, James Boswell referred to Webster as not learned but a "fund of information and entertainment...a very agreeable companion." In a recent book chapter describing Alexander Webster's contributions to the financial history of Edinburgh, the author and journalist Ray Perman writes that "Webster had married Mary Erskine, who had a fortune of £4,000 per year," which means that he had access to a sum far greater than his meagre wage. The contributions required by the scheme were relative miniscule for someone in his financial position.[60]

To our key point, Webster (obviously) selected the highest class (4th) when he joined at the age of 38 and made those contributions for almost 40 years. But his wife Mary—whom he married on 13 June 1737—predeceased him by 18 years, having died on 28 November 1766. So, ironically, upon the death of Reverend Alexander Webster on 25 January 1784, there was no widow's annuity to be paid.[61] One does not require any mathematical skill whatsoever to determine that Webster's internal rate of investment return was negative. Yet one doubts this financial fact would have disturbed him one iota. Why? In a letter sent to Webster during the summer of 1744, when ministers and professors were making their choices, the following complaint was lodged:

> Several old men are taking very undue advantages by claiming the benefit of the highest Class in the Widows Scheme; I have enclosed a letter which was insert in *The Evening Courant*, and beg you would improve the contents of it, with the ministers whose names are hereunto subjoined, to prevent their taking the highest class, for although the stock will rise whatever parts they act, yet it is really unjust for their heirs, to draw so much out of the fund, in virtue of their paying only a trifle.[62]

[59] Carlyle (1910, p. 1751).

[60] See Stewart quote from Boswell in ODNB entry. See Perman (2022, Chapter 6), and in particular the reference to the earlier Centisma Fund, created by the Scottish Episcopal Church in 1688. See also Dunlop (1992, Chapter 1), for a discussion of the (much) earlier scheme.

[61] See the *ODNB* entry noted above.

[62] This letter is stored in the National Records of Scotland, as CH9/17/32.

I'll return to this concern in the concluding chapter, but as readers can see, the unease expressed was that older contributors would be paying in "a trifle" since they would be expected to die very soon and quickly trigger the reversionary annuity for the wife and heirs. The writer—who pointed to a longer letter and article published in the newspaper—believed this to be unfair to younger participants who would be subsidizing the older ones. In fact, the above-noted letter in *The Evening Courant*, which this writer was alluding to, the author wanted to effectively "name and shame" the older participants who were unfairly selecting the highest class. In the terminology of Chapter 3, this was no fair longevity risk pool.

But Webster's response, who would have been around age 38 at the time and claiming to represent the views of the all younger brethren, wrote the following:

> Dear Sir. When you was out of town several of your brethren had a meeting, and it was judged of some consequence that each of us should write to our acquaintance among the younger brethren to the above purpose, pointing out such old men as they might apply to—and have sent you some copies of the printed letter to transmit to them—I have wrote to Mr. Turnbull to the same purpose. Mr. Cumming has dropped his pretensions to the Chair and the President has made up the matter. I am yours. A(lexander) W(ebster).[63]

It is unclear whether any of the older men changed their minds and choices, and the trustee minutes don't mention or reference this, but it seems the "younger brethren" were quite comfortable and at ease with the "old men" they were helping. Either way, the pretensions were dropped and the older men were (still) allowed to select the highest benefits. Before I conclude the matter and the chapter, I'll quote from the letter Colin Maclaurin had written a year earlier, in June 1743, when he provided his professional opinion about the "actuarial stability" of the scheme to Robert Wallace:

> For these and other reasons which it would be tedious to describe here at length, I sincerely wish well to this design; and cannot but be of opinion that if the scheme take place and be faithfully executed, as there is all the

[63] CH9/17/32.

> reason in the world to expect, it will prevent the unhappy circumstances to which minister's widows & children are too often reduced. It is remarkably advantageous to those ministers who are advanced in years, and they only seem to have any ground to complain of it who think they have no chance to leave a widow or child behind them. But as these are few in number, so it can only be said that they shall do for their brethren what their brethren must have done for them if it had been their lot to have had their circumstances exchanged.[64]

The solidarity expressed by Colin Maclaurin in this letter, and that expressed by Alexander Webster in his reply to a newspaper editorial, emphasized the sacrifice of brethren, who were clearly aware of the asymmetry of risk and financial circumstances and nevertheless endorsed the scheme. Perhaps that is one of the most important ingredients in these schemes, the glue that makes heterogenous longevity pools function: "They shall do for their brethren what their brethren must have done for them."

Conclusion

In addition to his work on the scheme, Alexander Webster is also known among Scottish historians (especially those from New Town, Edinburgh) as the author of the first attempted census, which was conducted in 1755, entitled appropriately enough "An Account of the Number of People of Scotland in the Year One Thousand Seven Hundred and Fifty Five," which also provided parish-by-parish population counts. His total for the entire population of Scotland in 1755/56, was precisely 1,265,380 people across 888 parishes for which data was available. Interestingly, a close examination of his census methodology, which was recently reviewed by the historian Michael Anderson and published in 2011, reveals a Webster consistent with his activities on behalf of the scheme for widows and orphans. To force parishes to return the population "lists" which he needed for the census, he threatened to withdraw their Christian school funding unless they sent them in on time. Perhaps this was a trick he learned while trying to collect those other lists. And, while some have questioned the "diligence in collecting, analysing and presenting" these figures, Anderson writes that "Webster often did not use exact numbers

[64] CH9/17/16.

that ministers submitted" and that "even for the eighteenth century, numbers in any document should not be treated as if they are accurate by modern standards." Nonetheless, Anderson concludes, the "account is an extraordinary achievement for which Webster should be commended." The only thing I might change is the phrase so it reads "should also be commended," as this was yet another of his contributions.[65]

A sermon was preached at Tolbooth Church of Edinburgh upon the death of Dr. Alexander Webster—he was granted a doctorate in divinity by the University of Edinburgh in 1760. The date of the sermon was noted in the *Scots Magazine* as having taken place on the Sabbath, 25 January 1784, and it published the following part of the eulogy:

> I now refer to the most ingenious useful and justly administered scheme, which the God of all wisdom and grace enabled him to contrive and accomplish. For the support of the widows and children of all the principals and professors of the four learned Universities in Scotland. This is a scheme of such invention, labour, excellence and real utility, as surpasses all praise. It will transmit the name of your late worthy pastor, with distinguished honour and respect, to the latest posterity. Suffice only to say of it in general, that it is founded on the best principles of calculation and a thorough knowledge of human life.[66]

At that point, his most notable contribution wasn't as census taker or as the 1753 moderator of the General Assembly of the Church of Scotland, but as the fund administrator. The unnamed eulogizer notes about Alexander Webster that "no haughty forbidding airs of superiority did he assume" and concludes with:

> During the time that it pleased the Sovereign Disposer of all events to allow its most ingenious author to remain with us, to superintend and direct its progress, it was put to such a full and satisfactory proof that under the care of providence of God it will be continue to be a sure and lasting fund

[65] See Anderson (2011), and in particular the quotes by R.A. Houston, T.C. Smout, and R.E. Tyson. Anderson pegs the population in 1755/56 between 1.15 and 1.35 million, compared against Webster's 1,265,380. See also the discussion around the use of Halley's life table, and a binomial test for randomness.

[66] Source: A Sketch of the Character of Alexander Webster, in The *Scots magazine* (1739–1803), February 1784, p. 61, via ProQuest.

of relief to the widows and fatherless for as long as the University shall flourish or the vestige of a Church shall be seen in our native land.[67]

Indeed, the 1744 Presbyterian scheme for widows and orphans lasted well into the late twentieth century. One can't help but wonder if the eighteenth century's Scottish literati's disdain for the evangelical Webster—a.k.a. Bonus Magnus, and unelected to the Select Society—somehow transported itself to scholars a quarter of a millennium later. His management methods might have been suspect, his numerical approximations questionable, and his drinking rowdy, but he got things done.

Sources and Additional References

Primary

Anonymous. (1784). Obituary: Sketch of the Character of Dr. Webster. *The Scots Magazine*.

Church of Scotland. General, Assembly. (1743). *A Scheme, for Providing an Annuity to Ministers Widows, and a Stock to Their Children; Humbly Offered to the Consideration of the Ministers of the Established Church of Scotland, by a Committee of the Commission of the Late General Assembly, Appointed to Draw up the Same, and Transmit It to the Several Presbyteries.* Printed by R. Fleming and A. Alison. Eighteenth Century Collections Online, https://link.gale.com/apps/doc/CB0126539754/ECCO?u=ed_itw&sid=bookmark-ECCO&xid=fcf599e7&pg=1. Accessed 12 July 2022.

Great Britain. (1744). *An Act for Raising and Establishing a Fund for a Provision for the Widows and Children of the Ministers of the Church of Scotland, and of the Heads, Principals, and Masters, of the Universities of Saint Andrew's, Glasgow, and Edinburgh.* Thomas Baskett and Robert Baskett. Eighteenth Century Collections Online, https://link.gale.com/apps/doc/CB0126539754/ECCO?u=ed_itw&sid=bookmark-ECCO&xid=fcf599e7&pg=1. Accessed 12 July 2022.

Great Britain. (1749). *An Act for explaining and amending an Act passed in the seventeenth year of His present Majesty's reign, intituled, An Act for raising and establishing a fund for a provision for the widows and children of the ministers of the Church of Scotland, and of the heads, principals, and masters of the Universities of Saint Andrews, Glasgow, and*

[67] Scots magazine, dated.

Edinburgh. Thomas Baskett and Robert Baskett. Eighteenth Century Collections Online, https://link.gale.com/apps/doc/CB0126539754/ECCO?u=ed_itw&sid=bookmark-ECCO&xid=fcf599e7&pg=1. Accessed 18 July 2022.

Halley, Edmond. (1693). "An Estimate of the Degrees of the Mortality of Mankind; Drawn from Curious Tables of the Births and Funerals at the City of Breslaw; with an Attempt to Ascertain the Price of Annuities Upon Lives." *Philosophical Transactions of the Royal Society of London, 17*(196): 596–610.

Price, Richard. (1771). *Observations on reversionary payments; on schemes for providing annuities for widows, and for persons in old age; on The Method of Calculating the Values of Assurances on Lives; and on The National Debt. To which are added, four essays On different Subjects in the Doctrine of Life-Annuities and Political Arithmetick.* 2nd ed. Printed for T. Cadell, in the Strand. Eighteenth Century Collections Online, https://link.gale.com/apps/doc/CB0126539754/ECCO?u=ed_itw&sid=bookmark-ECCO&xid=fcf599e7&pg=1. Accessed 12 July 2022.

Webster, Alexander. (1771). *A Letter from the Rev. Dr Webster, of Edinburgh, to the Rev. Dr Price, of London, and Dr Price's Answer, Relative to the Establishment for a Provision to the Widows and Children of the Ministers and Professors in Scotland.* Printed by Murray & Cochran. Eighteenth Century Collections Online, https://link.gale.com/apps/doc/CB0126539754/ECCO?u=ed_itw&sid=bookmark-ECCO&xid=fcf599e7&pg=1. Accessed 12 July 2022.

Webster, Alexander. (1740). *The Wicked Life, and Fatal but Deserved Death of Haman, Ahasuerus's Prime Minister. A sermon, preached before the Right Honourable, Lord Provost, Magistrates and Town-Council of Edinburgh, at their election, on Tuesday, 30th September, 1740.* Eighteenth Century Collections Online, https://link.gale.com/apps/doc/CB0126539754/ECCO?u=ed_itw&sid=bookmark-ECCO&xid=fcf599e7&pg=1. Accessed 12 July 2022.

National Records of Scotland (NRS) Church of Scotland (CH) Records

CH2/252/8: Lothian and Tweeddale synod, Minutes (1711–1721)

CH9/1: Church of Scotland Trustee Minutes; NRS Catalogue—catalogue record (nrscotland.gov.uk). https://catalogue.nrscotland.gov.uk/nrsonlinecatalogue/details.aspx?reference=CH9%2f1%2f13&st=1&tc=y&tl=n&tn=n&tp=n&k=&ko=a&r=ch9%2f1&ro=m&df=&dt=&di=y

CH9/3: Reports; NRS Catalogue—search results overview (nrscotland.gov.uk). https://catalogue.nrscotland.gov.uk/nrsonlinecatalogue/overview.aspx?st=1&tc=y&tl=n&tn=n&tp=n&k=&ko=a&r=ch9%2f3&ro=m&df=&dt=&di=y

CH9/4: Trustees' Scheme Books; NRS Catalogue—search results overview (nrscotland.gov.uk). https://catalogue.nrscotland.gov.uk/nrsonlinecatalogue/overview.aspx?st=1&tc=y&tl=n&tn=n&tp=n&k=&ko=a&r=ch9%2f3&ro=m&df=&dt=&di=y

- CH9/4/2: "Trustees' Scheme Books, 1744–1755."

CH9/5: Cash Books; NRS Catalogue—search results overview (nrscotland.gov.uk). https://catalogue.nrscotland.gov.uk/nrsonlinecatalogue/overview.aspx?st=1&tc=y&tl=n&tn=n&tp=n&k=&ko=a&r=ch9%2f3&ro=m&df=&dt=&di=y

CH9/6: Accounts, Cash Received and Paid; NRS Catalogue—search results overview (nrscotland.gov.uk). https://catalogue.nrscotland.gov.uk/nrsonlinecatalogue/overview.aspx?st=1&tc=y&tl=n&tn=n&tp=n&k=&ko=a&r=ch9%2f3&ro=m&df=&dt=&di=y

- CH9/6/2: "Accounts, Cash Received and Paid, 1747–1748."

CH9/7: Clerk's Ledgers; NRS Catalogue—search results overview (nrscotland.gov.uk). https://catalogue.nrscotland.gov.uk/nrsonlinecatalogue/overview.aspx?st=1&tc=y&tl=n&tn=n&tp=n&k=&ko=a&r=ch9%2f3&ro=m&df=&dt=&di=y

CH9/10: Schemes Re Charge of General Collector; NRS Catalogue—search results overview (nrscotland.gov.uk). https://catalogue.nrscotland.gov.uk/nrsonlinecatalogue/overview.aspx?st=1&tc=y&tl=n&tn=n&tp=n&k=&ko=a&r=ch9%2f3&ro=m&df=&dt=&di=y

CH9/12: Abstract of Facts and Remarks; NRS Catalogue—search results overview (nrscotland.gov.uk). https://catalogue.nrscotland.gov.uk/nrsonlinecatalogue/overview.aspx?st=1&tc=y&tl=n&tn=n&tp=n&k=&ko=a&r=ch9%2f3&ro=m&df=&dt=&di=y

CH9/15: Miscellaneous Volumes; NRS Catalogue—search results overview (nrscotland.gov.uk). https://catalogue.nrscotland.gov.uk/nrsonlinecatalogue/overview.aspx?st=1&tc=y&tl=n&tn=n&tp=n&k=&ko=a&r=ch9%2f3&ro=m&df=&dt=&di=y

CH9/16: Printed Reports; NRS Catalogue—search results overview (nrscotland.gov.uk). https://catalogue.nrscotland.gov.uk/nrsonlinecatalogue/overview.aspx?st=1&tc=y&tl=n&tn=n&tp=n&k=&ko=a&r=ch9%2f3&ro=m&df=&dt=&di=y

- CH9/16/1: "Printed Reports, 1744–1750."
- CH9/16/2: "Printed Reports, 1747–1790."
- CH9/16/3: "Printed Reports, 1751–1760".

CH9/17: Miscellaneous Documents; NRS Catalogue—search results overview (nrscotland.gov.uk). https://catalogue.nrscotland.gov.uk/nrsonlinecatalogue/overview.aspx?st=1&tc=y&tl=n&tn=n&tp=n&k=&ko=a&r=ch9%2f3&ro=m&df=&dt=&di=y

- CH9/17/1: "Representation and scheme for providing an annuity to ministers widows ... offered to the consideration of the ... Assembly of the church of Scotland to meet at Edinburgh the 12th May 1743.“ R. Fleming & Co., Edinburgh.

- CH9/17/2: "Scheme for providing an annuity to ministers widows, and a stock to their children; passed and resolved by the General Assembly ... 12 May 1743." R. Fleming & Co., Edinburgh.
- CH9/17/10 "Proposal for aising a fund for the widows of Ministers or others."
- CH9/17/11: "Calculations re Funds. Written by Dr. Wallace. 1743."
- CH9/17/16: "Letter from Colin Maclaurin to the Reverend Robert Wallace, 23 May 1743."
- CH9/17/23: "Letters from Committee of commission of General Assembly, Edinburgh, to Robert Wallace and George Wishart, at London, their commissioners sending them their opinions etc. 8 Dec 1743–15 Mar 1744."
- CH9/17/32: Letter and Reply by Alexander Webster
- CH9/17/34: "Calculations, with the principles and data on which they are instituted ... shewing the rise and progress of the fund." T. Lumisden & Co., Edinburgh, 1748.

Secondary

Ahnert, T. (2015). *The Moral Culture of the Scottish Enlightenment: 1690–1805.* Yale University Press.

Anderson, M. (2011). Guesses, Estimates and Adjustments: Webster's 1755 'Census' of Scotland Revisited Again. *Journal of Scottish Historical Studies, 31*(1), 26–45.

Carlyle, A. (1910). *The Autobiography of Dr. Alexander Carlyle of Inveresk, 1772–1805.*

Carmel, E. (2024) Moderation in the Scottish Enlightenment: The case of Robert Wallace. *History of European Ideas.*

Deuchar, D. (1896). Notes on Widows' Funds. *Transactions of the Actuarial Society of Edinburgh, 3*, 198–227.

Dow, J. B. (1975). Early Actuarial Work in 18th Century Scotland. *Transactions of the Faculty of Actuaries, 33*(240), 193–229.

Duncan, E. H., & Baird, R. M. (1977). Thomas Reid's Criticisms of Adam Smith's Theory of the Moral Sentiments. *Journal of the History of Ideas, 38*(3), 509–522.

Dunlop, A. I. (Ed.). (1992). *The Scottish Ministers Widows Fund: 1743–1993.* Saint Andrew Press.

Grabiner, J. V. (2002). Maclaurin and Newton. In C. W. J. Withers & P. Wood (Eds.), *Science and Medicine in the Scottish Enlightenment* (pp. 143–171). Tuckwell Press.

Garrett, A., & Harris, J. A. (Eds.). (2015). *Scottish Philosophy in the Eighteenth Century, Volume I: Morals, Politics, Art, Religion.* OUP Oxford.

Henderson, G. D. (1957). *The Burning Bush: Studies in Scottish Church History*. Saint Andrew Press.
Mackie, A. (1956). *Facile Princeps: The Story of the Beginning of Life Insurance in America*. Presbyterian Ministers' Fund.
Moore, J. (2008). Hutcheson, Francis (1694–1746), Moral Philosopher. *Oxford Dictionary*. https://doi.org/10.1093/ref:odnb/14273
Perman, R. (2022). *The Rise and Fall of the City of Money: A Financial History of Edinburgh*. Birlinn.
Phillipson (2010). *Adam Smith: An Enlightened Life*.
Riddell, L. K. (2021). Shetland's Population in the Eighteenth and Early Nineteenth Centuries. *Journal of Scottish Historical Studies, 41*(2), 105–133.
Stewart, M. (2004). Webster, Alexander (1707–1784), Church of Scotland minister. *Oxford Dictionary of National Biography*. Last Retrieved 28 November 2021. https://www.oxforddnb.com/view/10.1093/ref:odnb/9780198614128.001.0001/odnb-9780198614128-e-28939
Sageng, E. (2004). MacLaurin, Colin (1698–1746), mathematician and natural philosopher. *Oxford Dictionary of National Biography*. Last Retrieved 28 November 2021. https://www.oxforddnb.com/view/10.1093/ref:odnb/9780198614128.001.0001/odnb-9780198614128-e-17643
Sher, R. B. (2015). *Church and University in the Scottish Enlightenment* (2nd ed.). Edinburgh University Press.
Winch, D. (2007). Smith, Adam (bap. 1723, d. 1790), Moral Philosopher and Political Economist. *Oxford Dictionary of National Biography*, https://doi.org/10.1093/ref:odnb/25767
Wood, P. (2015). Thomas Reid and the Common Sense School. In Aaron Garrett & James A. Harris (Eds), *Scottish Philosophy in the Eighteenth Century, Volume I: Morals, Politics, Art, Religion*. History of Scottish Philosophy.

CHAPTER 7

Annuity Management in the Eighteenth Century

Was the Mathematics Known and Used?

David R. Bellhouse, a historian of mathematics, published a book in 2017 called *Life Contingent Contracts and the Emergence of Actuarial Science in Eighteenth-Century England*. The book discusses the development of actuarial science in the eighteenth century, specifically the mathematics behind the pricing and valuation of life annuities. Bellhouse highlights the fact that this fundamental mathematics was actually developed and published as early as 1725, which is 20 years prior to the establishment of the 1744 Presbyterian Fund for Widows in Scotland. It was Abraham De Moivre, a mathematician born in 1667, who published one of the earliest textbooks on valuation in 1725, building on the work done over a quarter of a century earlier by Edmond Halley.[1]

Despite the availability of this mathematical knowledge about the valuation of annuities and insurance, it did not significantly influence the insurance industry in the eighteenth century. According to Bellhouse, the influence on the trade was negligible until the end of the century. This observation is based on the work of another well-known historian of science, Lorraine Daston—whose work I will return to in Chapter 9. It is, however, interesting to note that although the mathematics for pricing and valuation of annuities was available in the early eighteenth century,

[1] Bellhouse (2017), Chapter 1.

M. A. Milevsky, *The Religious Roots of Longevity Risk Sharing*, https://doi.org/10.1007/978-3-031-62403-2_7

it took several decades before the insurance industry fully embraced the potential and impact of this knowledge in the pricing of insurance.[2]

In the present chapter, to begin, I think it necessary to mention a specific aspect of the annuity fund, which involves the technicalities associated with the fair valuation of widows' benefits. I intend to delve deeper into this topic in the subsequent section, where I will provide a set of formulas or equations for the value of the embedded life annuities. By doing so, it will lead to the fair payout multiple, which is another crucial aspect of this annuity fund.

I will assert that using these expressions is a valid approach within the eighteenth century because basic mathematics had already been developed by then. Therefore, the scientific community was well aware of the methodology, algorithms, and mathematical principles that underpinned the annuity fund's fair valuation. This assertion is significant because it demonstrates the robustness of the annuity fund valuation approach and its relevance in a historical context.

Theoretical Value: Payout Multiples

In this section, I will provide a detailed explanation of the mathematical calculations required to determine the value of the reversionary annuity, which was an essential component of the scheme for widows and orphans. Recall that the reversionary annuity is essentially a financial instrument that pays a guaranteed income to a beneficiary (the spouse, in the case of the 1744 fund) upon the death of the plan holder. To arrive at the value of the reversionary annuity, I need to factor in or assume various parameters such as the age of the plan holder, the contribution rates, and the mortality patterns of the underlying cohort. In particular, understanding the impact of age on contribution rates is crucial to determining the value of the annuity.

The key reason for going through all of this is that the argument I make in this section is that, in 1744, the scheme contributions were too low to fund the promised reversionary annuity. Therefore, the benefit multiple of 3.81 was too high, especially for the many older contributors. In other words, the scheme was not (initially) designed adequately. Note that the following paragraphs aim to provide a more in-depth insight into

[2] Daston (2008).

the mathematical principles used to value the reversionary annuity and its relevance to the scheme for widows and orphans. However, it is not meant to criticize or endorse the effectiveness of the scheme, nor suggest that its success or failure was solely due to mathematics.

While the economic forces of supply and demand generally drive prices, with mortality-contingent claims triggered by life and death, such as insurance and annuities, Adam Smith's invisible hand is subservient to matching *expected discounted actuarial values* (EDAV). The justification for using EDAV versus supply and demand is the so-called Law of Large Numbers (LLN).[3]

Longevity risk pooling—as alluded to already earlier a number of times in this book—refers to an approach to managing risk that is based on the idea that as the size of a pool of individuals increases, the uncertainty of outcomes associated with longevity risks declines. This is because the larger the pool, the more accurately the risks associated with the group can be predicted and managed.

In practical terms, this means that by pooling the longevity risks of a large group of individuals, such as through a pension fund or an insurance policy, the risks associated with individual outcomes are spread out across the group. This can help to mitigate the impact of unexpected events, such as an individual living longer than expected, on the overall group.

It is important to note that while the aggregate uncertainty of outcomes declines as the pool size increases, it is not possible to completely eliminate uncertainty. There will always be a degree of unpredictability associated with individual outcomes, especially in cases where the pool size is relatively small. However, by pooling the risks associated with longevity across a larger group, the potential impact of such uncertainty can be reduced.

Nevertheless, as many contributors are pooled together, the so-called idiosyncratic (a.k.a. non-systematic mortality) risk is diversified away, and economic valuation is by expectations (only). With that in mind, the reversionary annuity valuation process begins by defining the *net single premium* (NSP) for a standard life annuity as being equal to the above-noted EDAV of benefits. For the basic life annuity, the NSP is computed

[3] The earliest historical awareness of the law of large numbers can be traced to the Italian Gerolamo Cardano (b. 1501, d. 1576), and the first formal proof was by Jacob Bernoulli (b. 1655, d. 1705), in *Ars Conjectandi*, published posthumously in 1713.

using the following expression:

$$\overline{a}(r|x) := \int_0^\infty p(t|x)e^{-rt}dt \tag{7.1}$$

On the left-hand side of (Eq. 7.1), the first argument is the valuation rate (for example, $r = 4$ per cent) and the second denotes an initial age (for example, $x = 45$). On the right-hand side, the integral ranges from time zero to the end of the mortality table (that is, infinity.) Then, within the integrand, $p(t|x) \leq 1$, denotes the deterministic and known probability that a life currently age x will survive t years, and the interest factor: e^{-rt} discounts cash flows. Using the same logic, the NSP for an immediate life annuity that continues for the greater part of two separate lives (for example, husband and wife) is then:

$$\overline{a}(r|x, y) := \int_0^\infty p(t|x, y)e^{-rt}dt \tag{7.2}$$

The key difference between (Eq. 7.1) and (Eq. 7.2) is the survival probability $p(t|x, y)$, which represents the mathematical likelihood that either or both persons, of which one is currently aged x (for example, the husband) and another is currently aged y (for example, the wife) survives for t more years. The link between the two is via:

$$p(t|x, y) = 1 - (1 - p(t|x))(1 - p(t|y)) \tag{7.3}$$

Assuming a 30-year survival probability of 0.185 for a 45-year-old, it means that there is an 18.5% chance of reaching age 75. In other words, out of 100 people who are 45 years old, only 18 or 19 people can expect to live to age 75.

Now, let's consider two identical individuals of the same age. The probability that either one or both of them will not reach age 75 is 66.4%. This means that out of 100 pairs of identical people who are 45 years old, roughly 66 pairs will not reach age 75.

The key insight for the valuation of the reversionary annuity is that the annuity factor for the reversionary annuity can be thought of as the difference between two conventional annuities, one joint-life and one single. The reversionary annuity pays for the longer of two separate lives, but the reversionary annuity does not pay anything while the primary person currently aged is still alive. Using (hat) $\hat{a}$ for the reversionary versus (bar)

$\overline{a}$ leads to:

$$\hat{a}(r|x, y) = \overline{a}(r|x, y) - \overline{a}(r|x) \tag{7.4}$$

This applies only if the entire premium is paid *up front*. To arrive at the *periodic* contribution, (Eq. 7.4) must be amortized over the contributor's life. And, given the mandated nature of contributions, the denominator is a simple life annuity and not one that is extinguished at either the first of two deaths or the last of two deaths. Finally, to convert to a benefit multiple, one computes the inverse:

$$\text{Multiple} = \frac{\overline{a}(r|x)}{\overline{a}(r|x, y) - \overline{a}(r|x)} \tag{7.5}$$

The equations are the five key expressions for determining *theoretical* success or failure. The only step missing to provide the correct multiple in *practice*, is to calibrate $p(t|x)$, to a proper mortality table for the mid-eighteenth century in Scotland. This was done via a linear approximation to the (log) hazard rates from the Halley (1693) tables, also known as a Gompertz-Makeham approximation.[4] Again, the point of all of this is that I'm arguing that (at least initially) Alexander Webster and his co-managers didn't charge participants enough to cover the 3.81 benefit multiple, based on (7.5).

Now, at the time, the discussion and dialogue that took place around the proper structure of the scheme—for example, between Alexander Webster and Richard Price, or between Robert Wallace and Colin Maclaurin—didn't revolve around the proper application of these equations. Rather, the dialogue centred on how many new widows the fund could expect every year (around 18) or the total number of widows it

[4] The best fitting parameters were $\lambda = 0.01045$, $m = 68.05, b = 14.98$, in years. See the original article cited as Milevsky (2023) for more information on this approximation and tests for goodness of fit. Likewise, for similar attempts to value (old) annuities using (modern) techniques, see "Valuing Medieval Annuities: Were Corrodies Underpriced?" by Bell and Sutcliffe (2010). For more on the calibration of the Gompertz-Makeham model, see "Life and Death in Continuous Time", Milevsky (2020) or Tai and Noymer (2018). For more on Halley's life table, see "Edmond Halley's Life Table and Its Uses" in Ciecka (2008), as well as "A New Look at Halley's Life Table" in Bellhouse (2011), the collection of articles in Haberman and Sibbett (1995). For the first modern valuation (and variance) of the widow's annuity, see Boyle (1974). For the general pricing of longevity and mortality contingent claims, written by-and-for actuaries, see the textbook referenced as Dickson et al. (2009), in particular Chapter 5.

could expect to support (between 260 and 300). The point here is that the contributions in 1744 were too low.

Numerical Examples: Payout Multiples

Moving on to numerical examples for the reversionary annuity, let's consider a scenario where a 45-year-old contributor, such as a minister or professor, wishes to ensure that their 35-year-old spouse receives a lifelong benefit of £10,000 per year upon their death. This was the primary objective of the annuity fund, which charged annual contributions or premiums as outlined and explained in Chapters 5 and 6.

In theory this reversionary annuity could be acquired with one (large) payment of £44,200 at time zero—that is, when the primary contributor is 45 years old— which is the annuity factor in (Eq. 7.4) times £10,000. Of course, that is not how the annuity fund operated, since the payments were made annually as opposed to up front. Alternatively, then, the £44,200 can be divided or amortized by (Eq. 7.1), leading to £3650 per year while the contributor is alive (even if the spouse died first). That is closer to how the annuity fund operated.

Now, the ratio between the £10,000 annual benefit and £3650 annual contribution is a payout multiple of 2.74, so a £25,000 spousal benefit divided by 2.74 would cost £9125 per annum. The higher the payout multiple, the better the "deal" and the cheaper the benefits. Table 7.1, an amalgamation of our five key formulas as applied to the 1744 annuity, brings this all together. It not only computes the fair multiple but it also illustrates the important age dependency, where all numbers are predicated on the effective valuation (interest) rate of 4% per annum, continuously compounded.

Here is how to read Table 7.1 and interpret the various numbers displayed. An older contributor who is 55 years old and whose beneficiary is 35 years of age, would be required to contribute a fixed £6.08 per year—versus the £3.65 charged to ministers and professors in the 1744 annuity fund—to secure the £10 benefit for the widow. This is higher and more expensive than the above example (at age 35) because of the contributor's shorter (by 20 years) expected lifespan.

Notice the intuition here. If contributions or premiums that the contributor must pay while they are alive are amortized (that is, spread, paid) over shorter expected horizons, they must be higher to cover those benefits. Naturally, increasing the age of the spouse reduces the fair annual

Table 7.1 Theoretical payout multiples at 4% interest

Initial age of contributor	*Initial age of beneficiary*			
	30	*35*	*40*	*45*
30	4.55			
35	3.75	4.30		
40	3.03	3.47	4.05	
45	2.41	2.74	3.18	3.80
50	1.89	2.14	2.46	2.92
55	1.47	1.65	1.88	2.21
60	1.13	1.26	1.43	1.66
65	0.86	0.95	1.07	1.23

Note Based on Eqs. (7.1)–(7.5)

contribution. Why? An older spouse—all else being held constant—reduces the expected time over which the £10 annuity for life will be paid out, and therefore the expected discounted value of benefits is reduced, so the contributions can be lower. This follows from the logic of the five equations presented earlier.

In this context, let's consider the same table but a different cell. Suppose a contributor enters into an annuity agreement or a fund at the age of 30, and their spouse is also 30 years old. In this case, the contributor can secure a lifetime multiple of 4.55, which is quite high and more generous than what the 1744 scheme offered. This is because the contributor is expected to make payments for many years. On the other hand, if a 65-year-old with the same spouse (not exactly the same) enters the scheme, they can only obtain a lower multiple of 0.86, which is less than one. This is because they are not expected to be a part of the scheme long before they die. However, if the 65-year-old swaps their 30-year-old spouse for a 45-year-old, the appropriate benefits multiply jumps to 1.23. This is because the older spouse is not expected to draw benefits for very long if and when the contributor dies.

With all of this mathematics and numerical work out of the way, I am now able to provide a more definitive answer to the primary question: *Was the 3.81 multiple too high, generous, and unsustainable?* Well, if you look carefully at Table 7.1, you will notice that if everyone in the scheme had been exactly age 45 and had all their spouses been exactly 45 years of age at the time of entry into the fund, then the 3.8 multiples would have been suitable. In fact, it would have been perfect. Starting at the age of 45, every pound contributed would have entitled the beneficiary to a reversionary life annuity of 4.8 times that of one pound.

In fact, the same table numbers from Table 7.1 also tell us that for those ministers and professors who were younger than age 45 at the time they entered the scheme, the multiple of 3.8 would be "slightly unfavourable" because they should have been entitled to a higher multiple or payout. It certainly was sustainable if everyone was under the age of 45 at the time of entry.

However, this gets to the key insight from these last few pages of calculations and perhaps at the heart of why the fund terms had to be modified a few years later for those who were older (and there were plenty who were older in 1744): the scheme was simply too generous. The multiple should have been lower than 3.8, and this is why there was much discussion about the "older members" engaging in anti-selection against the fund. Moreover, the problem was magnified when these same older participants selected the larger benefits. All of these factors would inevitably lead to problems of sustainability, which leads us to the financial performance of the fund over time.[5]

The Annual Financial Statements

In this section, we will delve into the fund's financial statements. These statements were released annually in May, just before the General Assembly meeting of the Church of Scotland. Unlike modern financial statements that accompany a corporation's annual report, the fund trustees released a single combined statement, which was a hybrid of an income statement and a statement of cash flows. It is important to understand the difference between an income statement and a cash flow statement. An income statement summarizes a company's revenues and expenses over a specific period, showing net profit or loss. It follows the accrual accounting principle, meaning it records revenues when earned

[5] Echoing what was mentioned earlier in this chapter with respect to the work of Bellhouse (2017), the importance of age-dependent pricing for life annuities was known to the early eighteenth-century mathematicians, such as Abraham de Moivre (b. 1667, d. 1754) and Edmond Halley (b. 1656, d. 1742), but was not accounted for in the structure of the fund terms.

and expenses when incurred, regardless of cash movement. This statement gives insights into the company's operational efficiency and profitability.

On the other hand, a statement of cash flows details the actual cash inflows and outflows from operating, investing, and financing activities over a specific period. It highlights how a company generates and uses cash, providing a clearer view of its liquidity and financial health. This statement is crucial for understanding the company's ability to generate cash, pay debts, and fund operations. While the income statement can show profitability, the cash flow statement reveals whether a company can sustain its operations and grow over time, making it essential for assessing its financial viability. Well, in the case of the annuity fund, the trustees released a document or report that combined those two, which I'll denote as the Charge & Discharge (C&D) statement, which is what the trustees appeared to have labelled it. The C&D statement is reminiscent of a two-sided balance sheet, with offsetting entries, but in this case the two sides show money going into the hands of the collector, and money coming out. The C&D statement makes absolutely no mention of the assets of the fund, and certainly no reference—or even a hint—to the discounted liabilities of the scheme. Again, all that was reported on the left-hand side was the amount of money the General Collector accumulated into the fund in that year from contributions, interest, and maturing loans, and the amount of money that flowed out of the fund in the form of benefits and expenses. In theory, the two sides of the C&D statement should balance—or at least be extremely close to each other. The small difference between these two sides, which could be either positive or negative, was referred to as the collector's balance. Table 7.2 provides a summary of the C&D statement during the first decade.

Let's review the columns from left to right. Every year the General Collector would receive (1) minister and professor annual contributions, (2) interest earned on the loans of the assets of the fund, and (3) maturing loans. Those first three columns represent the charge. Then, moving to the right we have the discharges.

First, there is payout of management fees (to himself), the issuing of new loans, and then the payment of benefits. Those are columns 4 to 6 in Table 7.2. There is one other entry or column, and that was meant to capture the difference between any arrears that had either not been paid or not been claimed in the prior year, which was settled in the current year. I know this might start to sound complicated, and yet it is critical for appreciating how the fund was managed, so here is a detailed example.

Table 7.2 Fund charge and discharge: 1745 to 1755

Year	*+Annual cont.*	*+Interest earned*	*+Old loans maturing*	*−Mgmt. Fees*	*−New loans made*	*−Benefits paid*	*−Net Arrears*	*= Collectors Balance*
1745	£8068	£14	£ –	£234	£7680	£ –	+£334	£19
1746	£5279	£215	£150	£129	£3570	£531	+£841	£591
1747	£4779	£377	£420	£241	£3480	£2126	-£87	£409
1748	£4933	£515	£447	£243	£3525	£2382	-£208	£363
1749	£4956	£628	£509	£249	£3660	£1892	+£194	£462
1750	£4957	£722	£868	£249	£4050	£1522	+£641	£547
1751	£5027	£890	£480	£249	£3034	£2013	-£80	£1731
1752	£5081	£1034	£990	£249	£5940	£2617	+£194	-£160
1753	£4980	£1302	£1320	£249	£3013	£2460	+£628	£1092
1754	£5069	£1393	£1350	£249	£4070	£3335	+£268	£982
1755	£5025	£1525	£1350	£249	£3560	£4398	+£86	£761
Total	£58,154	£8615	£7884	£2590	£45,582	£23,276	-£2639	

Source Annual Reports, 1745–1791. NRS 9/16. Only pounds reported

Sample "Charge and Discharge"

I'm going to select the year (row) 1750 in Table 7.2, and describe exactly how the final collector's balance of £547 was computed, starting with the prior (1749) year's ending collector's balance, which is the opening (1750) collector's balance of £462. First, in the year 1750, the annual contributions made were £4957, according to the lists sent in by the presbyteries for that year. Then, interest earned from the loans made out by the fund amounted to £722. The old loan balances that matured and were paid back by the contributors in 1750s amounted to £868. This results in a total charged (money in) balance of £6547, which is simply: 4957 + 722 + 868. That's relatively easy to understand and account for, and clearly, most of the loans didn't mature (or were not paid back) in any one given year. Now let's examine the discharges.

First, the fees paid to management, specifically the clerk and the collector in 1750, amounted to £249, which was a standard and an agreed-upon amount. Notice how, other than in the year 1746, that discharge or outflow for management fees was the same in pounds, although not as a per cent of the fund, unlike modern management expense ratios. Moving on, the new loans made amounted to £4050 in the year 1750. The annuity benefits paid to widows and children for their deceased family members were reported as £1522. This number could be confirmed or audited by checking the lists of widows. Finally, the net arrears for amounts that were due in prior periods, such as loans that were to be paid out or contributions to be paid in, is £641 for the year 1750. This brings the total discharge (money out) for the year 1750 to £6462, which is the linear sum of the four prior numbers, 249 + 4050 + 1522 + 641. Finally, the collector's ending balance is equal to the beginning balance plus all the charges, minus all the discharges, which is: £462 + £6547 – £6462 = £547. Check the last column for 1750, and everything balances. One can do the same from the top of the table (in 1745) to the bottom (in 1750). All the money going in or coming out was accounted for.

Narrative Versus Numerics

Similar to how they are constructed today, each annual report was presented as follows. First there was the opening statements, which noted the ministers and professors who had died within the year. Then, there was a listing of all those who were entitled to benefits, organized by women and children. Then, there were additional comments, followed by financial forecasts and projections for the future state of the fund—these near prophecies are something I'll get to in a moment—then a discussion of the accuracy of prior predictions, and finally, the financial statements (Charge & Discharge) themselves. Indeed, not unlike modern annual reports, the narrative, the comments and the footnotes around the financial statements revealed just as much if not more than the numbers did. Ever since corporate financial statements have been compiled, likely in the year 1603 when the Dutch East India Company, known as *Vereenigde Oostindische Compagnie* (VOC) in Dutch, was chartered, the narrative around the statements has been just as important as the numerics within the statements. And, while VOC was disbanded and ceased trading in the late eighteenth century, in fact it was technically bankrupt, the Presbyterian annuity fund in contrast lasted well into the late twentieth century.[6]

[6] Note. The company was bankrupt by 1799. See the thesis by Petram (2011), on the world first stock traded on the world's first stock exchange.

Garbage-in-Garbage-out

Students of accounting are routinely warned that the validity and accuracy of the high-level reports they compile for management and shareholders are dependent on the quality of the data that went into generating those reports. This is often summarized in textbooks as the maxim that "garbage in equals garbage out" and we can see an early precursor of this principle in the reports accompanying the fund's financial statements.[7] In a 1745 letter to Alexander Webster, written by Alex Bryce, which appears on page 33 in the document with a shelfmark of CH9/16/2, he writes about some of the "garbage" that might be going into the calculations. Bryce, in the preface to the financial statements, expresses that he knows Alexander Webster will be unsatisfied with the basis of those statements and acknowledges that there are inaccuracies in them, partly due to the "*Unhappy situation of the country*" and partly due to the presbyteries not drafting up proper lists as requested. Needless to say, the unhappy situation was the so-called Scottish elephant in the room, the Jacobite invasion of 1745, which I discussed in Chapter 6.

Another interesting note, in the same letter, was a claim that ministers were complaining about the collector because he was demanding interest for their annual rates when they were not duly paid on time. The point here is that if the General Collector was complaining about the quality of the lists which were supposed to be compiled by the ministers, and yet they in turn were complaining about his conduct, this must have created a very unhappy situation (and dynamic) indeed.

Recall that the General Collector for the fund and his clerk during the first few years of the operation of the fund were James Stewart (collector) and Robert Macintosh (clerk). Interestingly, in the aftermath of the above-noted unhappiness, namely, in the year 1746, both of their salaries or fees (noted in Table 7.2) were halved,[8] which, although not mentioned or explained explicitly or anywhere within the notes, does speak to some of the financial discretion in managing the fund, something I alluded to in the previous chapter. And, to this exact point, in 1749 the trustees expressed concerns regarding the (new) estimates that the fund would have to support 400 widows, when the original plan called for only

[7] See, for example, the textbook Horngren et al. (2018).

[8] The source for this 50% reduction in fees during the year 1746, is CH9/16/2, pg. 43.

334. At that point there was explicit mention in the annual reports that management fees would have to be reduced and the budget for the (new) building might have to be cut.[9]

Speaking of management fees for the General Collector and his clerk, and being fully aware that this is a sensitive topic in the twenty-first century as it was in the eighteenth, there was one other (interesting) source of revenue for the collector: the interest he charged to the fund. For example, in the year 1752, the General Collector's balance (the last column in Table 7.2) dropped to –£160. That is a negative number, and represents a deficit that came out of the collector's pocket. But for that, he was compensated with interest of £23, which was charged to the fund.[10]

"Must Rise to This Amount"

The overall impression from reviewing the annual reports, written by the fund trustees to the members of the General Assembly of the Church of Scotland, is one of managing public relations as well as managing fund financials. In fact, it wasn't uncommon to see a preamble to the financials—which were distributed widely—that were at odds with the (more private) meeting minutes, noted in Chapter 6. For example, in the opening statements of the annual report for 1747, the committee announced that they were happier with the record-keeping of the fund, yet the message from the meeting minutes during the exact same time appeared to tell a different story.[11] A few years later, in 1750, the committee appears to have done the exact opposite, when they complained that "Not received any lists from the said presbytery there two years past."[12] But the most interesting takeaway was the sense that the financial affairs of the fund—whenever there was any discretion in the matter—were managed towards the original budget and projections. This was noted and emphasized in Chapter 6 where the (more) private minutes were transcribed and discussed, but even make an appearance in the publicly distributed reports. In one particular passage from the

[9] CH9/16/2, pg. 45.

[10] For this, see particular example see: CH9/16/3, pg. 39A and CH9/16/3, pg. 53A.

[11] CH9/1/1.

[12] CH9/16/3, pg. 11.

report,[13] dated 9 May 1751, one that could even be described as a smoking gun, the trustees explicitly note that certain decisions must be taken because the fund "must rise to this amount." Which amount was that? The original budgets. They were targets and not forecasts.

To understand how the fund values were computed from the statements of charges and discharges, I'll present a numerical example for one of those years.

From Charges & Discharges to Fund Values

Here is the algorithm or the formula for the stock calculation, which oddly enough only appears in the report for the second year of the fund.[14] The calculation begins with the prior year's free stock balance, adds the charge balance, including the collector's balance, then subtracts the beginning collector's balance, subtracts the previous year's arears, subtracts repaid bonds, subtracts management fees and finally subtracts the benefits paid. This might all seem rather abstract and convoluted, so here is a detailed example for year 1750 of the fund, which is one of the years that I examined in detail for Table 7.2.

Step 1: The free stock balance in 1749 amounted to £21,946 per the 1749 financial statements. Step 2: The charge balance in, including last year's collector balance, amounted to £8083, which is: 462 + 7621. Step 3: Subtract the beginning collector's balance of £462. Step 4: Subtract the previous year's arrears due of £1074. Step 5: Subtract old loans maturing of £868. Step 6: Subtract the management fees of £249. Step 7: Subtract benefits paid to widows and children £1522. In sum, the ending stock balance (calculated) is: 21,946 + 8083 – 462 – 1074 – 868 – 249 – 1522 = £25,854, and the actual ending balance, per the financial statements, is: £25,837. This is a difference of £17.

"Money Lying Dead"

To this point, the only interest rate noted or mentioned has been the 4%, but as time went by, the trustees wanted to charge more. It's unclear whether British interest rates increased during this time, which were

[13] CH9/16/2, pg. 97.

[14] CH9/16/2, pg. 45.

generally in the range of 4% to 5% during the eighteenth century.[15] The trustees debated increasing the interest rate charged on the £30 loan to ministers, and more generally—as the size of the fund increased—the rate charged to merchants who borrowed from the fund. This excerpt speaks to increasing the rate from 4 to 5% on loans that were past due (i.e., late):

> Upon report of the Committee touching the period from which five pounds per cent interest is to be charged for the thirty pounds lent upon bond, this meeting do direct that the Collector do charge upon such bonds as are repayable, interest at five per cent from the term of Whitsunday or Martinmas immediately preceding the time of the promotion, translation, deprivation, resignation or death of the grantor of such bond.[16]

In fact, as the size of the fund grew over time—and was projected to continue growing (per Table 7.3)—the trustees were faced with a decision on what to do with funds that were not being lent out to ministers and professors. After all, and to keep things simple, if there were 1000 participants and each borrowed £30 (in theory), then what was to happen in the mid-1750s, when the "stock" of the fund, which is £30,000 (in theory), exceeded the supply of ministers and professors who could borrow the assets?[17] Yes, some of the cash holdings, above and beyond the ongoing income, were used to pay benefits and annuities, but there would clearly be more than what the ministers and professors could absorb in loans. Well, one proposal was to force everyone to borrow more, and this speaks to that idea:

> A fourth proposal relating to the clause in the Act of Parliament...appointing thirty pounds to be lent to each contributor, should be extended to fifty pounds under the same conditions...The Trustees delay the same 'til the necessity thereof appear.[18]

[15] For more on this, see the article by Weiller and Mirowski (1990). Research by Chamley (2011), seems to indicate that during the two decades in question, risk-free British interest rates were declining from 4 to 3%.

[16] NRS, CH9/1/1, pg. 295, meeting 31, dated: 12 January 1748.

[17] I selected the £30,000 deliberately, as this was the sum originally approved as the maximum value of the fund's equity, and it was then increased in a later revision to £50,000. If the equity exceed those amounts, the "surplus" was to be proportionally distributed to the scheme's current orphans and widows.

[18] NRS, CH9/1/1, pg. 301, meeting 33, dated: 8 March, 1748.

Other ideas around what could be done with the excess "stock" of the fund that was not lent out to contributors, and wasn't needed to pay required benefits, was to distribute excesses to beneficiaries. Here is one excerpt that speaks to that, and the objections raised:

> A ninth proposal about dividing the surplus arising in any one year (after the capital of thirty- five thousand pounds shall be made up) among the widows and children of contributors, in proportion to the annual rates of their deceased husbands or fathers, being read, is agreed to...The provision of extending the capital is only designed to take place in case of necessity, as the Church has no plot to have any larger fund than what is absolutely needful.[19]

In the end, the decision was taken to lend the extra assets of the fund to bankers, merchants, entrepreneurs, and even the city of Edinburgh itself. The process of lending money to those entities—which today might be called private loan placements, or holding illiquid debt—was done with extreme diligence and care. Think of a modern-day (bank) loan underwriter "kicking the tires" of the business and especially the assets securing the loan before offering a farthing. The following discussion speaks to this matter:

Table 7.3 Scheme fund projections vs. actuals

Year	*Projections in 1748*	*Actual amount*	*Difference £*	*Diff. %*
1750	£26,092	£25,837	–£255	–1.0%
1755	£41,110	£40,750	–£360	–0.9%
1760	£51,012	£50,164	–£847	–1.7%
1765	£58,348	£58,347	–£1	0.0%
1770	£63,963	£65,732	+£1769	+2.8%
1775	£68,761	£72,328	+£3567	+5.2%
1780	£73,470	£77,133	+£3662	+5.0%
1785	£77,943	£83,758	+£5815	+7.5%
1790	£81,759	£89,090	+£7330	+9.0%

Source Annual Report, 1791. NRS 9/16. Only pounds reported

[19] CH9/1/1, NRS, pg. 338, meeting 36, dated: 22 September 1748.

> Upon report of the Committee that after the loans of thirty pounds to each Minister and Professor in Scotland, there will be money to lend out to other hands...the Trustees appoint Professor Gowdie, Masters Robert Kinloch, Robert Wallace, Alexander Webster, John Erskine and Kenneth Mackenzie as a Committee to consider the proper method of laying out and securing the money of the Fund so to be lent, and they are appointed to have their first meeting on Monday next at five o'clock afternoon in this place, with power to them to adjourn to such times and places as they shall see proper, and to report to next stated meeting of the Trustees.[20]

Of course, Alexander Webster was front-and-centre in this initiative as well, as a subcommittee, together with the second member of the original triumvirate Robert Wallace, since by then Colin MacLaurin was deceased. The first proposal they considered was from the City of Edinburgh.

> The Committee appointed to converse with the magistrates and Town Council of—Edinburgh about lending to the town part of their growing capital stock, reported that pursuant to such appointment they have had frequent meetings with the magistrates and that the town are willing to borrow any sum from the Trustees not exceeding twenty thousand pounds sterling under the following terms and conditions and not otherwise.[21]

They also had discussions with other borrowers, and, in particular, they considered a proposal from Sir Thomas Kirkpatrick. This is what he offered:

> The Trustees, having received a proposal from Sir Thomas Kirkpatrick of Closeburn for borrowing at Whitsunday next four thousand pounds sterling of the capital stock at the rate of four and an half per cent. of interest on heritable security upon lands to the extent of three hundred and fifty pounds sterling of yearly rent free of all deductions and incumbrances each year's annual rent of the said principal sum to be paid within three months after due, otherwise the legal interest to be payable for that year; did unanimously agree to the said proposal; and the Trustees further considering the loss which the Fund sustains by money lying dead in the General Collector's hands happening to be paid betwixt terms, did also agree to

[20] CH9/1/1, NRS, pg. 396, meeting 47, dated: 12 February 1751.

[21] CH9/1/1, NRS, pg. 425, meeting 52, dated: 10 December 1751.

> authorise the Committee to lodge these sums at such interest as shall be agreed upon in the hands of Messrs.[22]

The above quote is from a meeting that took place in mid-January (1752), and Whitsunday would have been months away (in June), so something had to be done with the funds for those sixth months, and the trustees wanted to ensure the money would not "lie dead" in the interim.

As the fund grew and the money had to be reinvested, the sums lent out increased, and Alexander Webster drove some hard bargains and extracted even higher interest rates, including trigger payments and penalty rates for late interest. This is from a meeting in February 1753, presumably with the same concerns about dead money sitting in the hands of the collector.

> Mr Webster represented to the Trustees that Mr John Swinton advocate, had proposed to the committee to borrow from the Trustees a sum to the extent of six or seven thousand pounds sterling of the capital stock at the rate of four and a half per cent. Of interest on heritable security, that if the Trustees could not lend him the whole of the said sum at Whitsunday next, he was willing to take what they could give at that term and would take the rest afterwards as the Trustees could allow him to have it, and that each year's annual rent of the sum so borrowed should be paid within three months after falling due, otherwise the legal interest at five per cent. should be paid for that year.[23]

Finally, it's quite remarkable to read the extent of the due diligence and care in ensuring the borrowers had sufficient property and income to pay the interest. In the eighteenth century (and well afterwards) the true measure of wealth wasn't the value of the property pledged as security, which few bothered calculating, but rather the rental income that the property or pledged asset generated. At no point did anyone attempt to discount future cash flows, arrive at an NPV, or mark the asset to market. Rather, this is the procedure used to determine whether a borrower or his collateral was worthy of a loan:

[22] CH9/1/1, NRS, pg. 434, meeting 54, 14 January 1752.

[23] CH9/1/1, NRS, pg. 463, meeting 60, dated: 20 February 1753.

> ....Opinion of Messrs. Crawford and Mackenzie as to Mr Swinton's proposal...Having considered a proposal made by Mr John Swinton, advocate, to the Trustees of the Fund for a provision to Ministers' widows in Scotland for borrowing of them six or seven thousand pounds sterling of their capital stock upon heritable security on his lands and estate of Swinton and having considered the titles under which the said Mr John Swinton possesses his estate, with the preceding memorial drawn by the Clerk to the Trustees anent the incumbrances affecting the same and having also seen a rental given in by the said Mr Swinton and signed by him and Mr Henry Veatch, Minister at Swinton, whereby it appears that the gross rent of said estate amounts to eight hundred and sixty pounds, fifteen shillings and seven pence sterling, and of which it is informed the public burden at an eight months cess are thirty-three pounds, seventeen shillings and eight pence sterling....we are humbly of opinion that the Trustees may safely lend to Mr Swinton seven thousand pounds sterling of their capital stock on the security offered by him viz.[24]

So, in sum, the funds could be "safely" loaned to Mr. Swinton, and another £7000 of the annuity scheme's assets could be allocated to fixed income.

Minister's Arbitrage Versus Usury

In addition to their own attempt to extract higher yields from the assets, trustees of the scheme knew that some participants who took out loans from the fund at 4% interest were turning around and lending out the loans at 5%; effectively engaging in interest rate arbitrage. This guaranteed 1% spread that ministers and professors earned on a loan of £30 would net them 6 shillings per year,[25] a non-trivial supplement to their benefices and salaries. This practice wasn't an isolated or inconsequential aspect of the scheme, and the minutes described this process as one in which ministers "*got a hand to whom he may lend it.*"[26]

[24] CH9/1/1, NRS, pg. 469–470, meeting 62, dated: 22 May 1753.

[25] The 30 pounds is equivalent to 7200 pence, so 1% interest is 72 pence per year, which divided by 12 (pence per shilling) is 6 shillings in interest, although whether it was risk-free depended on whether any creditors defaulted.

[26] CH9/1/1.

With all this discussion about interest rates, be they 4% or 5%, individually or collectively, some readers might wonder how the eighteenth-century Church of Scotland could openly condone charging interest, and by ministers of all people. These ministers and fund trustees would no doubt have committed to memory the biblical text in Deuteronomy, Chapter 23, Verses 19–20, which states:

> You shall not charge interest on loans to your brother, interest on money, interest on food, interest on anything that is lent for interest. You may charge a foreigner interest, but you may not charge your brother interest, that the Lord your God may bless you in all that you undertake in the land that you are entering to take possession of it.

It's unlikely these ministers were lending their £30 allotment to foreigners, and the text from Deuteronomy is but one of many biblical admonitions against interest and usury. Recall that Reverend Alexander Webster was well known for his fire-and-brimstone Sunday sermons. How could this practice—deep immersion in the intricate mechanics of charging interest—have been condoned by Protestant ministers?

Well, the answer might be found in the writings of the French theologian John Calvin, who is most responsible for the teachings of the Reformed tradition of Protestant Christianity, including the doctrine of predestination, which I'll get back to in Chapter 9 when I discuss *probabilistic beliefs*. John Calvin was a contemporary of John Knox and greatly influenced Knox's thinking. In fact, they both lived together in Geneva in the late 1550s, when Knox was forced to leave Scotland (by Mary of Guise) for political reasons.[27]

To the matter of charging interest, Calvin broke from the Roman Catholic's unambiguous condemnation of usury. The thirteenth-century theologian Thomas Aquinas, in *Summa Theologica*, argued that charging interest was against natural law, and this remained the Roman Catholic view well into the sixteenth century. But in Protestant countries such as England and the Netherlands, the stigma was fading, and John Calvin's writings (arguably) were the catalyst or at least the conduit for change.[28]

Now, this isn't the venue for debating the extent to which Calvin (or Calvinism) condoned or was ambivalent about the charging of interest.

[27] Dawson (2015).

[28] George, C.H. (1957).

The prevailing view was that according to Protestants, interest and usury should be left to civil and not ecclesiastical courts. Calvin's letter *De usuris*, and his commentary on the relevant verses in the Book of Exodus—limiting any interest prohibitions towards the poor—aren't a clear-cut endorsement of money lending. Nevertheless, Calvin's writings on interest were far from Aquinas or the Aristotelian condemnation of usury, even if the distance is nebulous. Despite the noted ambiguity, John Calvin was asked about the specific practice at the heart of this chapter: ministers lending money to fellow congregants. Presumably, the 1744 fund was lending to Scotsmen who were members of some congregation. Well, it appears this was a common enough practice that Calvin (nearly two centuries earlier, of course) was asked to opine on the matter. He responded very cautiously as follows, in a letter dated 10 January 1562:

> I should most probably have been more expeditious in giving you an answer…had I not felt some scruples in touching on so delicate a subject, that of ministers lending money upon interest.[29]

At first, Calvin appears to hedge and tries to avoid responding all together:

> I dare not assert that it is not lawful. But on the other hand, when I consider to how many calumnies and scandals such a practice may lead, and also that many persons are apt to pass the bounds of moderation in following it, and think themselves warranted to make such profits as may be deemed illicit, I willingly abstain from giving any answer to this question.[30]

Instead of citing Deuteronomy or Exodus, his guidance is from the later prophets in the Hebrew Bible:

> The safest and most expedient conduct would be not to engage in such practices or contracts, and it is not without reason that Jeremiah

[29] Letters of John Calvin, Volume 4, page 252, Internet Archive, Contributed by University of Toronto, available here: https://archive.org/details/lettersofjohncal04calvuoft/page/252/mode/2up.

[30] Ibid, page 253, top.

protests that the contestations in which he was involved arose neither from borrowing nor lending.[31]

Jeremiah, of course, was protesting during the late sixth and early fifth century BC, just as Babylon was about to destroy the city of Jerusalem. But, after hinting at both sides, the final dispensation by John Calvin moved away from the prophet Jeremiah, and was pragmatic and affirmative:

> As this practice is more supportable than pursuing mercantile speculations, or conducting any traffic by which he might be diverted from his functions, I see no reason why the thing should be condemned in general. Nevertheless, I could wish that so much moderation were observed that people should not desire to derive a certain profit from it, but should content themselves with lending their money to some merchant, a man of integrity, and trust to his good faith and loyalty for making an equitable gain out of it, when God should cause his industry to thrive.

There we have it, almost two centuries before the Presbyterian ministers in Scotland moonlighted as loan arbitragers, and the trustee of the scheme hunted for advantageous interest rate opportunities, John Calvin granted the theological backing for this exact practice; as long as it was done with moderation. Arguably, the 1% spread from charging 5% interest to debtors and paying 4% to creditors fell under Calvin's sense of moderation. To be very clear, though, this practice of dealing in interest wasn't limited to the fund for widows and orphans and became widespread across all the financial affairs of the Church of Scotland during the eighteenth century. Most recently this was discussed by Mutch (2017), in an article published in the *Scottish Historical Review*, where he noted the conflicting dynamics between church practice and religious belief.

Conclusion and Takeaways

What are we to take away from these very old financial statements and activities from over 280 years ago? Do they tell us anything important about the origins of longevity risk sharing? Well, I believe the answer is yes. In particular, the narrative around the financial statements supports

[31] Ibid, page 253, bottom.

my thesis that the scheme's longevity and success were not only due to the mathematical calculations and projections made by Maclaurin in the early 1740s but also to the management of the fund itself, especially over the next decade. Moreover, the fund was managed with those calculations as budgets, a roadmap of sorts, a north star, as opposed to a perfect mathematical prediction that came true. Perhaps the trustee viewed the original projections like a biblical prophecy that had to come true. Either way, this process involved much discretion. Recall the quote by the trustees, that the fund "must rise to this amount," which evokes a self-fulfilling prophecy.

So, perhaps like a template that is used to guide spending, anywhere the trustees had some discretion, they exercised it in the direction that would guide the fund towards the original "prophecy." If the fund value was coming in short of the target, they would (threaten to) cut down on expenses or deny claims. And, as the years went by and the fund's value increased, the level of strictness was reduced—which was also evident from the meetings minutes reviewed in Chapter 6. Recall that all the actuarial calculations were based on a 4% rate, and the trustees endeavoured to charge 5%. As to why it was so important to meet the original expectations, it could have been a matter of instilling public confidence in the scheme, or perhaps—and yes this is speculative—an attempt to heap additional praise on the founders of the 1744 church scheme.

Sources and Additional References

Primary

Halley, E. (1693). An Estimate of the Degrees of the Mortality of Mankind; Drawn from Curious Tables of the Births and Funerals at the City of Breslaw; with an Attempt to Ascertain the Price of Annuities upon Lives. *Philosophical Transactions of the Royal Society of London, 17*(196), 596–610.

Price, R. (1771). *Observations on Reversionary Payments; On Schemes for Providing Annuities for Widows, and for Persons in Old Age; On the Method of Calculating the Values of Assurances on Lives; and on the National Debt. To Which Are Added, Four Essays on Different Subjects in the Doctrine of Life-annuities and Political Arithmetic* (2nd ed.). Printed for T. Cadell, in the Strand. Eighteenth Century Collections Online. http://link.gale.com/apps/doc/CW0105622537/ECCO?u=ed_itw&sid=bookmark-ECCO&xid=2190f221&pg=1. Accessed 12 July 2022.

National Records of Scotland (NRS) Church of Scotland (CH) Records.

CH9/1: Church of Scotland Trustee Minutes; NRS Catalogue—Catalogue Record. nrscotland.gov.uk

CH9/16: Printed Reports; NRS Catalogue—Search Results Overview. nrscotland.gov.uk

- CH9/16/2: Printed Reports, 1747–1790.
- CH9/16/3: Printed Reports, 1751–1760.

Secondary

Bell, A., & Sutcliffe, C. (2010). Valuing Medieval Annuities: Were Corrodies Underpriced? *Explorations in Economic History, 47*(2), 142–157.

Bellhouse, D. R. (2011). A New Look at Halley's Life Table. *Journal of the Royal Statistical Society: Series A, 174*(3), 823–832.

Bellhouse, D. R. (2017). *Leases for Lives: Life Contingent Contracts and the Emergence of Actuarial Science in Eighteenth-century England.* Cambridge University Press.

Boyle, P. P. (1974). Note on the Variance of a Widow's Pension. *Journal of the Institute of Actuaries, 101*(1), 103–107.

Chamley, C. (2011). Interest Reductions in the Politico-financial Nexus of Eighteenth-century England. *The Journal of Economic History, 71*(3), 555–589.

Ciecka, J. E. (2008). Edmond Halley's Life Table and Its Uses. *Journal of Legal Economics, 15*(1), 65–74.

Daston, L. (2008). Classical Probability in the Enlightenment. In *Classical Probability in the Enlightenment* (New Edition in 2023). Princeton University Press.

Dawson, J. (2015). *John Knox*. Yale University Press.

Dickson, D. C. M., Hardy, M. R., & Waters, H. R. (2009). *Actuarial Mathematics for Life Contingent Risks.* Cambridge University Press.

George, C. H. (1957). English Calvinist Opinion on Usury: 1600–1640. *Journal of the History of Ideas, 18*, 455–474.

Haberman, S., & Sibbett, T. (1995). *History of Actuarial Science* (Vols. 1–10). Pickering & Chatto.

Horngren, C. T., Datar, S. M., Foster, G., Rajan, M., Ittner, C., & Gowing, M. (2018). *Horngren's Cost Accounting: A Managerial Emphasis* (9th Canadian ed.). Pearson Education Canada.

Milevsky, M. A. (2020). Life and Death in Continuous Time. *Retirement Income Recipes in R: From Ruin Probabilities to Intelligent Drawdowns* (pp. 131–157). Springer-Verlag.

Milevsky, M. A. (2023). Adam Smith's Reversionary Annuity: Money's Worth, Default Options and Auto Enrollment. *Financial History Review, 30*(2), 162–197.

Mutch, A. (2017). The Business of Religion: Lending and the Church of Scotland in the Eighteenth Century. *Journal of Scottish Historical Studies, 37*(2), 136–154.

Petram, L. O. (2011). *The World's First Stock Exchange: How the Amsterdam Market for Dutch East India Company Shares Became a Modern Securities Market, 1602–1700* (Doctoral dissertation, Universiteit van Amsterdam [Host]).

Tai, T. H., & Noymer, A. (2018). Models for Estimating Empirical Gompertz Mortality: With an Application to Evolution of the Gompertzian Slope. *Population Ecology, 60*(1–2), 171–184. https://doi.org/10.1007/s10144-018-0609-6

Weiller, K. J., & Mirowski, P. (1990). Rates of Interest in 18th Century England. *Explorations in Economic History, 27*(1), 1–28.

CHAPTER 8

From Church PAYGO to Fully Funded

Introduction and Objectives

The prior chapters have focused on the details of the 1744 scheme, the ongoing management of the fund by the trustees, and the Enlightenment personalities who contributed to the fund over the first decade. In this chapter, I go back in time even further. The point of this excursion is to help address the as-of-yet unanswered questions of *how and why* the scheme emerged and succeeded at that particular point in time. The argument I will advance in this chapter is that the intellectual origins and administrative roots of the 1744 scheme can be traced back over a quarter of a century earlier, to a micro-scheme created and approved by the Church of Scotland in May 1718. The earlier design, which I'll discuss, has been long forgotten and was flawed from an actuarial or financial perspective. Nevertheless, it's quite possible that a more rapid organic development of the (eventual) 1744 scheme was stunted by preoccupations within the Church of Scotland during the first few decades of the eighteenth century.[1]

[1] Dunlop (1992, p. 7) briefly notes earlier attempts, but is silent on the details or the church context, which is the focus on this chapter.

M. A. Milevsky, *The Religious Roots of Longevity Risk Sharing*,
https://doi.org/10.1007/978-3-031-62403-2_8

To understand this argument, and that the origin of the world's first successful longevity risk pooling scheme had earlier roots within the Church of Scotland, or that it didn't emerge whole cloth in a burst of scientific and actuarial brilliance in the early 1740s, it will be necessary to delve into some of the minutia of Scottish and British history during the preceding 2–3 decades. This was a period bookmarked on both ends by the first British prime minister, Robert Walpole, who was in office from 1721 to 1742, and who "fell" just a year prior to the formal proposal in 1743. Is this just a coincidence? I'll have more to say about Robert Walpole and the relevance of his tenure to the scheme, but first, the background.

To begin, a recent article by Aaron Allen, which was published in the *Scottish Historical Review* in 2022, discusses the history of retirement pension arrangements by and for craft workers, going as far back as the 1670s. In that article he notes that annuity fund schemes—such as the 1744 scheme for Church of Scotland ministers—"replaced or at least augmented discretionary pensions." Notice the word replaced, suggesting that there was something prior and the finished plan evolved from something else. The contribution of the (final) funded schemes, versus prior ad hoc arrangements is that they offered greater dependability and reliability. Allen also notes that while the popularity of schemes that offered annuities for widows increased in the eighteenth and nineteenth centuries, not all such annuity schemes were obligatory on all members. So, in that sense the 1744 Presbyterian scheme—which was obligatory on all members—was different, more reliable, and hence successful. Again, there were earlier attempts, even within the Church of Scotland, and that is what this chapter is about.[2] As readers will learn, early attempts to fund life annuities are reminiscent of the modern debate around the transition from *pay-as-you-go* (PAYGO) pension *arrangements* to funded schemes.[3]

[2] By earlier attempts, I am referring to developments going back decades, versus a few years prior to the scheme at the heart of this book. For example, an early 1742 proposal, which both Alexander Webster and Robert Wallace objected to in principle, had elements of a *pay-as-you-go* scheme. In fact, some of that early (paygo mentality) DNA remained in the bones of the 1744 scheme, until those unfunded aspects were removed by the 1748 amendments.

[3] Allen, Aaron (2022), 'Corporate Charity for "the House": Craft Pensions and the Widows' Fund, 1670–1782', *Scottish Historical Review* 101, pp. 210–247.

Fasti Ecclesiae Scoticanae

I have heaped much praise on the triumvirate of Alexander Webster, Colin Maclaurin, and Robert Wallace and the many other diligent trustees of the 1744 scheme. In Chapter 6, I quoted from the 1784 eulogy for Alexander Webster, both of which were quite effusive about his contributions to the scheme and the Church of Scotland more generally. Oddly enough, though, within the pages and documents describing the history of the many other ministers of the Church of Scotland over the years, a slightly different picture emerges. Credit appears to be distributed more widely than previously thought.

For researchers interested in Scottish church history, a very important collection of books is known as the *Fasti Ecclesiae Scoticanae*. This collection is referenced as Scott (1915) in the bibliography, but often has other editors associated with the different volumes. The Latin title often accompanies the subtitle: The Succession of Ministers in the Church of Scotland from the Reformation. To save some space, the shorter abbreviation *Fasti* is the common name given to this collection, which at its essence contains a short historical biography of every single minister from the Church of Scotland.

On page 119 of Volume 1 of *Fasti*, which covers the history of the Synod of Lothian and Tweeddale, there is a full-page description of our protagonist Alexander Webster. Normally the biographies are much shorter, but Webster's is quite long. Unsurprisingly, *Fasti* notes that on 15 May 1744, he received the thanks of the General Assembly of the Church of Scotland, "for the extraordinary pains and trouble taken by him in the rise and progress of the scheme," which was noted as the first of its kind in the country. The same biographical entry in *Fasti* also notes that in June 1771—which is well after the temporal limits of this book—Webster was made General Collector of the fund, the individual who was responsible for the charges & discharges, as discussed in Chapter 7. *Fasti* helps clear up some questions around authorship of unnamed documents, as it credits Webster as the author of the 1748 publication entitled *Calculations with the Principles and Data on which they are Institute, relative to the Widow's Scheme*, which is one of the most important primary sources for the scheme.

Oddly enough, the same *Fasti* collection has a separate entry on page 60 of the same volume for another minister from Edinburgh whose name was John Mathieson (b. 1679, d. 1752). Reverend Mathieson was ordained to the church in Tolbooth (Edinburgh) in December 1706, and does appear in the meeting minutes of the trustees over 30 times before his death in 1752. Given his age and location, one would have expected him to be involved with the scheme and, indeed, according to the archival lists, he selected the highest £25 reversionary annuity for his wife. This was quite fortuitous, as you might have realized from his date of death in 1752, which means he only contributed for six or seven years before he died. As a senior minister, Mathieson was also the acting *praesis* [*sic*], which is the chair or convenor, in one of the early meetings in June 1746. The archival records also show that he diligently borrowed the £30 from the fund, in January 1745, and duly paid his 4% interest to the fund.

So far, Reverend Mathieson sounds like many of the other trustees and church "movers and shakers," albeit not on the magnitude of Alexander Webster. Not that name counts are the only metric for relative importance, Webster's name appears nearly a hundred times in the minutes, while during the same period and prior to his death, Mathieson appears approximately 30 times. But here is where it gets interesting. The *Fasti* bibliographic entry for John Mathieson states two very surprising facts. First, it notes, "He is said to have suggested in 1741, the establishment of a fund for supporting the widows and children of the ministers of the Church." According to *Fasti*, it was Mathieson's idea. So, while Webster, Wallace, and Maclaurin get the fame in the modern telling—see the usual Harari (2014) or Ferguson (2008)** references, for example—an unsung founder appears to be the Reverend John Mathieson. OK, you might be thinking to yourself, "So there was someone else, more senior, who first gave the idea for the 1744 scheme, and he has been ignored." Alas, how many people through history—and especially women—have been forgotten for their achievements? In fact, the so-called Stigler Law of Eponymy, named after the economist Stephen Stigler, states that no scientific discovery is named after its original discoverer. Stigler himself noted this as an example of the phenomenon, as he attributed the concept to sociologist Robert K. Merton, who had earlier written about processes through which scientific discoveries are named. I'll return to Robert K. Merton in the next chapter when I discuss science and religion. But, as for naming "things" after their true originator, the oversight might be due to historical, cultural, or linguistic reasons. In other words, you think you know who invented or suggested or created something—because of the name—and more often than not it's wrong.

But here is where it gets even more intriguing, and the reason I have digressed so far. In the subsequent section of the above-noted biography of Reverend Mathieson, the *Fasti* author also refers to someone else in addition to Mathieson creating a fund for widows and orphans: "That may be true so far as the fund established in 1744 was concerned, but a similar proposal has been made as early as *1716 by Patrick Couper, minister of Pittenweem*" (emphasis my own). When I first read this one sentence line, I couldn't help but wonder, who exactly was Reverend Couper and what was his proposal—25 years earlier—all about? To me it felt like being on an archaeological dig for a year, excavating an ancient Roman city, and then uncovering an even older Greek one underneath.[4]

Patrick Couper, Pittenweem, and Witches

First off, Pittenweem is a charming fishing village located in the East Neuk of Fife, Scotland. It is along the coastline of the Firth of Forth, and today Pittenweem is known for its pretty harbour, historic buildings, and lively arts scene. The odd sounding name Pittenweem comes from the old Pictish and Gaelic languages, meaning "Place of the Caves." That reference is to the cave-like storerooms which were and still are used by fishermen at the harbour. Less humorously, among the village's sordid history is that during the late sixteenth and early seventeenth centuries, it was one of the Scottish towns that put on witch trials, in which many poor innocent men and women were sentenced to death. And, get ready for it, the Reverend Patrick Couper, noted above as the originator of the earliest attempts to create a proper fund for poor widows and orphans, was the minister in Pittenweem during the witch trials. In those records, the reverend's name often appears as Cowper, or Cooper, but it's the same person. One account of the witch trials notes: "The local minister, Patrick Cooper, was on a mission to expose alleged witches, obsessed with uncovering the Devil's work."[5]

[4] See the entry for John Mathieson in Scott (1915), St Giles, p. 60.

[5] For the witch trials, the entry on Historic UK website (https://www.historic-uk.com/HistoryUK/HistoryofScotland/The-Pittenweem-Witch-Trials/). The above quote is from the article published in Historic Environment Scotland (https://blog.historicenvironment.scot/2023/03/pittenweem-witch-hunt/), in March 2023, entitled: *Murder, Malice and*

Interestingly, while the entry for John Mathieson gives the earliest credit for a scheme to Patrick Couper, the *Fasti* entry for the devil-obsessed reverend makes no mention of annuities or witches or much else. Not much is known about Patrick Couper, who also spelled his name as Cooper or Cowper. His entry in *Fasti* notes that he was ordained in a meeting house in Bannockburn in 1688 and that he was transferred to his ministry in Pittenweem, in May 1692, where it seems he remained for almost 50 years. He arrived in Pittenweem a decade before the witch trials he oversaw, and over two decades before he made his 1716 proposal, so he must have been a rather senior church figure. I find no mention of Patrick Couper having survived to, or participated in, the 1744 scheme; he must have been long gone from the church and scene by then.[6] A record from the Annals of Pittenweem indicates that he died in 1740, when another minister took his place, proving that he lived into his mid-70s and would have been active in the church for a fairly long time. It's all rather odd. The Mathieson entry in *Fasti* deliberately takes credit away and gives the original idea to Couper, but the *Fasti* entry for Couper mentions absolutely nothing meaningful about him. Was he a devil-chasing zealot or annuity innovator? Perhaps he was both. Indeed, one is almost tempted to dismiss Reverend Patrick—and the existence of an earlier Church of Scotland annuity scheme—were it not for the fact that the proposal itself is historically accurate and well documented. Here is the rest of the story.[7]

Misogyny. The Pittenweem Witch Hunt, written by Alea Ibrahim. See the following 2012 article in the Scotsman newspaper (https://www.scotsman.com/news/villagers-vote-on-memorial-for-witches-1639618), for a description of the proposed memorial. See also Henderson, Lizanne (2016), *Witchcraft and Folk Belief in the Age of Enlightenment Scotland, 1670–1740*, Palgrave MacMillan, ISBN 978-1-137-31324-9.

[6] See Scott (1915), p. 314. Note the 1744 scheme has David Couper participating.

[7] Dunlop (1992), p. 7, notes the difficulty in properly assessing credit for the idea, briefly mentions Cowper (versus Couper), notes his early 1716 proposal, that he was dead by 1740, but there is no mention of his witch hunting activities.

The Earlier Scheme of 1717

The minutes of the General Assembly of the Church of Scotland for May 1717 include a detailed discussion of a proposal to arrange life annuities for the widows of ministers. Not only was it proposed but a full and proper Act was passed by the General Assembly in 1718, enshrining the plan. The discussions and outline of the plan included the following stipulation[8]:

> ...every Minister in Scotland, to give and dedicate the tenth part of his stipend for one year for the end foresaid and do appoint the same to be paid in money and no[t] otherwise to the moderator of each Presbytery...The Assembly declares that his relict and children shall be entitled to a share of the fund in that synod where the said minister paid...[t]he money collected be turned into a stock and that it be lent out upon good security and that only the yearly interest thereof be distributed and that the stock be kept entire. And, if the annual rent shall at any time happen to do more than serve the exigencies of the Widows and Children in the synod, they appoint the said surplus annual rent to be added to the stock.[9]

OK. Let's look at the numbers. The *tenth part* noted above—that is a contribution rate of 10%, not a fixed amount—is reminiscent of the "teinds" or tithes on all forms of income directed to the parish churches. The General Assembly of 1718 was quite clear that these micro or mini schemes were to be locally managed without cross-subsidy or any spreading of benefits across synods.

> The contributors in every Synod shall have the management of what sums shall be collected and given for the ends foresaid to the widows and children of ministers who have served and contributed within their bounds and that none other shall have rights thereto.[10]

Let me be very clear. This wasn't just an idea, a plan, or a proposal. It seems the early scheme was implemented, and individual synods themselves took charge of the matter. No, this was not a large, centralized fund,

[8] During this period Between 1722 and 1739, 304 wives of Church of Scotland ministered were widowed. See also CH9/17/1, and CH9/17/2.

[9] Acts of the General Assembly, 14 May 1718, p. 6.

[10] Ibid., p. 7.

like the 1744 scheme, but a local synod-by-synod scheme, which could be managed by individual presbyteries. To clarify some of the church terminology here, think of a large federal fund, versus a smaller provincial or state arranged fund, versus an even smaller city plan. That was the relationship between the General Assembly of the Church of Scotland (federal) versus the synods (provincial) versus the presbyteries (city). The 1718 scheme was to be managed locally, individually, with little in the way of sharing or pooling risk. The only decision to be made was whether the pool should be very small (by presbytery) or slightly larger (by synod).

The reference to this plan goes beyond an Act of the General Assembly. The minutes of the Synods of Lothian and Tweeddale for November 1718 indicate a recommendation to the several presbyteries to consider whether to "keep and manage themselves or join with the synod" in the management of a fund for widows and orphans. Again, the General Assembly allowed management by the individual presbyteries or synod as long as the tenth (10% of the wage or salary or benefice) was contributed to the fund. In the Synod of Lothian and Tweeddale, in May of 1719, the meeting minutes indicate the following decisions: The Presbytery of Dalkeith reported they had a fund of their own and "according to the allowance of said act" would continue to manage it on their own. The Presbytery of Dunbar similarly reported that they wanted to manage it themselves. In contrast, the presbyteries of Biggar and Peebles decided to join a larger synod-based fund. And, the presbyteries of Linlithgow, Hadington, and Edinburgh hedged their replies. They resolved to "come to terms" and collect the tenth as required by the Act of the General Assembly of 1718, but they didn't commit to joining the synod fund.

Perhaps one of the issues they struggled with was the spirit of Chapter 31 of the *Westminster Confession of Faith*, which states: "Synods and councils are to handle or conclude nothing but that which is ecclesiastical and are not to intermeddle with civil affairs."[11] An argument can be made that collecting and paying out life annuities—and the financial money management it entailed—was not ecclesiastical in nature. Regardless of the motives here, the smaller 1718 schemes—the Couper schemes—were not the Webster, Wallace, Maclaurin (or Mathieson) longevity risk pools of 1744. The best way to describe the precursor and earlier annuity schemes would be as longevity risk ponds or even

[11] CH2/252/8, p. 354.

puddles. I use those terms because the synod and presbytery schemes were unsuccessful for a variety of reasons, including the fact they were closer to pay-as-you-go schemes instead of funded schemes that attempted to actuarially project future liabilities, create reserves for those liabilities, and set contributions to finance those reserves. In that sense, 1744 was one small textual step in the annals of the Church of Scotland, but an enormous actuarial step forward for mankind.

From Synod Versus Presbytery

For fear this might be confusing, and to place the lack of efficacy of these earlier schemes in perspective, here is an overview of the size of synods versus presbyteries. How big were they? These are figures presented by the Scottish historian Alasdair Raffe who is a Professor at the University of Edinburgh. He notes that during the period 1660 to 1714 there were 67 (local) presbyteries across Scotland, which in turn were organized into 13 synods. This is an average of approximately five presbyteries in each synod.[12] And, even more roughly speaking, this leads to an average of 70 ministers in each synod scheme, and obviously less in each presbytery-based scheme. My statistical point here is that this number was not large enough to diversify longevity risk. It was indeed a pond or a puddle.

And, here is one final financial point about the early Couper scheme. The one-time, lump sum payment of one-tenth an annual stipend, on the order of £6 pounds—using the average Church of Scotland stipend—for a total fund size of £420 was woefully inadequate and it would certainly not generate stable life annuities for widows either. If 4% interest on £420 generated £17 per year, this might be enough to cover just one widow, and the pond and puddle funds would quickly evaporate as the number of widows increased. Nevertheless, perhaps the genesis of longevity risk pooling began with a failure. Namely, an early unsuccessful attempt to do

[12] Cited as Raffe (2012), *The Culture of Controversy: Religious Arguments in Scotland 1660–1714*, p. 2. The names of synods were 1. Aberdeen, 2. Dumfries, 3. Glasgow and Ayr, 4. Merse and Teviotdale, 5. Sutherland and Caithness, 6. Angus and Mearns, 7. Fife, 8. Lothian and Tweeddale, 9. Perth and Stirling, 10. Argyll, 11. Galloway, 12. Moray, and 13. Ross. Also, To be clear, this is a very rough average since presbyteries differed in size.

the same thing—with the wrong price and the wrong methodology—in the second decade of the eighteenth century.[13]

Daniel Defoe's Path to a Financial Innovation

Widening the temporal scope of the analysis, the early eighteenth century was a period of "financial revolution" with many proposals for social and economic improvements or "projects" being floated. In fact, one of the best-known promotors for annuity schemes was none other than Daniel Defoe (b. 1660, d. 1731). Defoe was a writer, trader, and even a spy, who today is best known for his novel *Robinson Crusoe*. He was born with a more common name, Foe, and later added "De" to his surname to sound more impressive. In addition to detailing the travails of Robinson Crusoe, he wrote several other novels and essays, and was a financial entrepreneur, or at least he tried to be. Over the course of his life Defoe faced financial difficulties, including bankruptcy and even imprisonment for debt.

In a 1697 collection of essays simply titled *Essay Upon Projects*, Defoe proposes a government entity called the Office of Insurance for Widows, echoing the scheme at the heart of this article. Here was the essence of his plan:

> We have an abundance of women...with nothing to support them, which falls generally upon the wives of the inferior clergy, or of shopkeepers and artificers...If he dies, then she is absolutely undone, unless she has friends to go to...Suppose an office to be erected, to be called an office of insurance for widows...Two thousand women, or their husbands for them, enter their names into a register to be kept for that purpose, with the names, age, and trade of their husbands, with the place of their abode, paying at the time of their entering 5s. down with 1s. 4d. per quarter...If any one of the women become a widow at any time after six months from the date of her subscription...she shall receive the sum of £500 in money without any deductions.[14]

Defoe's plans came to naught (and his pricing was widely off mark) perhaps because of the trauma caused by the South Sea bubble in the

[13] The sums are speculative, and it's also unclear whether all synods implemented the micro-scheme.

[14] Defoe (1697), "An Essay Upon Projects", see Section II—For Widows.

1720s, but it is interesting to note that a few decades later the Church of Scotland implemented something similar. Indeed, given his relationship with the church—and time in Scotland—during the period of the Union of 1707, an argument could be made that Defoe, a Presbyterian, might have (also) sowed the idea for the annuity fund in Scotland.

Andrew McDiarmid positions the Scottish financial revolution as taking place decades after the English financial revolution, a view echoed above by Defoe. McDiarmid differentiates between the initial phase of the Scottish revolution coinciding with the English one in the late 1690s, but he claims that in Scotland the financial revolution was stymied by the failure of the Darien scheme and the upheaval around the Acts of Union in 1707. McDiarmid argues that the second and third decades of the eighteenth century "denoted a phase of the revolution in which financially innovative projects returned to [Scotland] after a period of absence." So, perhaps schemes in the late 1690s, such as those suggested by Defoe, made their way north from England (and from the Continent) in the second decade of the eighteenth century.[15]

To be very clear, the 1718 Church of Scotland's earlier micro-scheme was incomplete, fragmented, and unfunded. The records of the General Assembly in the lead-up to the 1744 scheme express unhappiness with the local plans. So, quite justifiably, Colin Maclaurin, Robert Wallace, and Alexander Webster deserve credit for (1) creating a proper fund, and (2) it on a sound mathematical basis. But they did not create the idea out of whole cloth. Alexander Webster would have been 10 years old and both Colin Maclaurin and Robert Wallace in their late teens in 1716, when Minister Patrick Couper, made his original suggestion that has been all but forgotten.

[15] McDiarmid (2021), 'The Equivalent Societies of Edinburgh and London, the Formation of the Royal Bank of Scotland, and the Nature of the Scottish Financial Revolution', p. 89. See also Watt (2007), *The Price of Scotland: Darien, Union and the Wealth of Nations*, and Sussman (2022) 'Financial Developments in London in the Seventeenth Century: The Financial Revolution Revisited', for the scope of the financial revolution.

From Small to Large

The idea of unifying synod-based funds into one larger fund to be managed centrally was vigorously debated and then summarily dismissed in 1735. One gets the sense of a philosophical struggle between advocates of national versus local schemes. Here is how they expressed it in 1743:

> The General Assembly having considered a scheme proposed for one general fund…also considering the scheme for synodical funds… judging that said scheme for synodical funds should be adhered to and vigorously advanced as being free of several great difficulties which would necessarily occur both in raising and managing of the more general fund. There being now no prospect of any success in the said scheme for a general fund and whereas the project for the said general fund has for some years past very much interrupted the usual and ordinary contributions to the synodical funds which are by the above Act of Assembly 1718 required to be paid by every minister.[16]

The "several great difficulties" could have been referring to the difficulty in collecting the funds and/or the lack of scale to help spread the risk. Either way, the question is why, after 25 years, did widows' annuities go from micro-unfunded to macro-funded. It could be that the poor performance of the synod- and presbytery-based funds convinced the General Assembly they weren't viable. Alternatively, perhaps the impetus was part of a larger struggle between localized control of the synod's financial affairs versus a centralized (Edinburgh-based) national church. To examine that angle, let's examine what was happening during that quarter-century.

Church Turmoil

To begin with, the year 1716 was a few years after the passage of the highly contentious Patronage Act of 1712 ("A Great Grievance"), an event with too strong a gravitational force to be ignored in any discussion of the period. In fact, it's impossible to overemphasize the impact of patronage on developments within the Church of Scotland over the

[16] Church of Scotland, General Assembly (1743), *A Scheme, for Providing an Annuity to Ministers Widows …*, p. 7.

decades leading up to the 1744 annuity fund, and it wasn't a clear-cut case of for or against.[17]

In brief, the Patronage Act of 1712, was a piece of legislation passed by the Parliament of Great Britain. The main provisions were as follows. First, the Act restored the rights of patrons to present (think of it as nominating) ministers within the Church of Scotland. This right had been abolished by the Presbyterian party in 1690 after the Glorious Revolution of 1688. The 1712 Act meant that landowners and landed gentry who were the patrons of parishes could (again) appoint ministers to vacant churches. A power transfer of sorts, and it was highly controversial and very widely opposed within the rank-and-file public in Scotland. At the time, the Presbyterian establishment wanted democratically elected ministers of the congregations, and the 1712 Act was seen as an imposition by the English Parliament on Scottish religious affairs. This so-called restoration of patronage rights contributed to the formation of secessionist churches, because congregations resented the imposition of ministers by patrons. The point here is the 1712 Act—that is just a few years before the first proposal—represents a critical moment in the history of the Church of Scotland, reflecting the tensions between religious independence and state authority. That was the ambiance, the background, and the context, and historians of the period have noted the long-term impact.

The Church of Scotland experienced a significant schism in the nineteenth century, known as the Disruption of 1843, and the division was primarily over this exact issue of patronage. The Disruption culminated in 1843 when a large group within the Church of Scotland left to form the Free Church of Scotland, advocating for the principle of spiritual independence and congregational election of ministers. While this is well after the historical scope of the 1744 scheme, this split resulted in about one-third of the clergy and a substantial number of laity leaving the established church, which resulted in the creation of a parallel system of Free Church institutions. Moreover, the fund was split as well, until the churches were brought back together in a 1929 reconciliation.[18]

[17] Whitley (2013), *A Great Grievance: Ecclesiastical Lay Patronage in Scotland until 1750*, p. 127; Raleigh (1921), *Annals of the Church in Scotland*, p. 290; Buchan (2007), *Capital of the Mind: How Edinburgh Changed the World*, p. 63; and Sher (2015), pp. 24, 48.

[18] See Dunlop (1992) for more on the split and how it affected the fund.

Returning to the eighteenth century, historians Richard Sher and Alexander Murdoch write that "patronage struggles…were considerably more complex than is usually thought"—and more complicated than the standard ecclesiastical histories of Scotland would suggest. They document that by the year 1750 the number of secession churches numbered around 45 congregations. Of course, those ministers would not have been part of the 1744 scheme, and the potential pool was shrinking.[19]

The point here is that this was a turbulent time in the church. Moving to the 1720s, the General Assembly was preoccupied with a controversy known as "The 12 Marrow Men" and the republication of *Marrow of Modern Divinity* by Edward Fisher. The Church of Scotland condemned the book (and Fisher) because—among other rather abstract and technical theological points in the book—it devalued holiness as necessary for salvation.[20] By the year 1722, the General Assembly was busy condemning Ralph and Ebenezer Erskine.[21] From 1726 to 1729, the church was pursuing the University of Glasgow professor John Simson for heresy. Why? In the year 1714, the church took issue with Professor Simson's views on the theological matter of free will versus predestination. In the late 1720s, the church's concern with John Simson was regarding his (possible) Arian views that Christ was "less fully divine than God the Father." All of that turmoil was a prelude to Ebenezer Erskine and the Secession in the 1730s, which originated with the General Assembly's attempt to place the franchise for ministerial elections in the hands of heritors (owners of heritable property) and elders.[22]

My point here is not to get lost in the minutia of the Church of Scotland's history, but rather to ask why a proposal or idea to create a scheme for widows in 1716 took 20 twenty years to implement successfully. Answer: Those two decades were busy ones.

[19] Sher and Murdoch (1983), p. 169.

[20] Whitley (2013), p. 153, as well as Van Doodewaard (2009), 'The Marrow Controversy and Seceder Tradition: Marrow Theology in the Associate Presbytery and Associate Synod Secession Churches of Scotland (1733–1799)'. See also Kidd (2016), 'The Fergusson Affair: Calvinism and Dissimulation in the Scottish Enlightenment', for more on church affairs between 1720 and 1740.

[21] Coincidentally, Ebenezer was the brother-in-law of Alexander Webster, the scheme's champion.

[22] Whitley (2013), p. 186, as well as the extensive discussion in Sefton (1962).

This conflict between Archibald Campbell (Earl of Ilay) and Ebenezer Erskine and his supporters lasted until 1736, at which point yet another crisis erupted. That was the year of the infamous Porteous riots in Edinburgh, in which Captain Porteous was condemned to death for firing on a riotous crowd. His death sentence was (controversially) absolved by Queen Caroline, leading to further riots and the eventual lynching of Porteous by an Edinburgh mob. Some clergy supported Porteous (and the Queen) while others favoured the Edinburgh mobs—and more turmoil.[23]

This is a bird's-eye view of the turbulence encapsulated by the statement "the clergy have been so divided for some years past."[24] By early 1741, matters might have sufficiently calmed so other initiatives could be advanced. What might have happened in the absence of these agitations—or whether the fund would have emerged earlier—is obviously counterfactual. The point is that the emergence of the scheme in 1744 can't be detached from church history.[25]

To this point, the historians Sher and Murdoch write that while the 1740s were a "hot" decade for patronage presentation disputes, they were an extremely "cold" one for general crisis.[26] The Earl of Ilay was replaced by the Marquess of Tweeddale and made Secretary of State for Scotland in March 1742. Robert Wallace, one of the triumvirate, who wasn't on good terms with either Ilay or Robert Walpole, the prime minister, might have been waiting for a political opening during a "cold" period in the Church of Scotland.[27]

[23] Whitley (2013), p. 211, and perhaps the initial chapters of Sir Walter Scott's novel *The Heart of Midlothian*.

[24] In a letter from Lord Advocate Robert Craigie to the Marquess of Tweeddale, see Smith (1973), 'Clerical Corridors of Power: Extracts from Letters Concerning Robert Wallace's Involvement in Ecclesiastical Politics, 1742–1743'.

[25] See, again, Whitley (2013), Chapter 12.

[26] Sher and Murdoch (1983), p. 273.

[27] Per Smith (1973), as stated in a letter dated 5 May 1743 from Thomas Hay, 'He [Wallace] will be chosen and moderator by a great majority and will have all Mr. Webster's interest' which links Wallace and Webster early on.

As Britain's first de facto prime minister, Robert Walpole served from 1721 to 1742, making him the longest-serving holder of that office in history. Walpole's fall from power in 1742, perhaps giving an opening to Robert Wallace to propose legislation for the annuity scheme, was precipitated by a combination of factors. His long tenure bred complacency and arrogance, leading to increased opposition. The military failure of the war against Spain and the consequent unpopularity eroded his support, which eventually forced Walpole to resign. To historians, Walpole's resignation marked the end of an era, and a year later the world's first successful longevity risk pooling scheme was born in Scotland.

To Sum Up the Controversy and Timeline

The summary in Table 8.1 provides a list of the key dates and events within the Church of Scotland, from the original proposal in 1716 to the final launch of the scheme in 1745. One cannot help but notice how the dates are also aligned with the consensus on the temporal limits of the Edinburgh-based Scottish Enlightenment. The Moderates Hugh Blair and William Robertson—as part of their responsibilities in Edinburgh—were initial trustees of the annuity scheme, and the time was ripe for an Enlightenment scheme that might help unite disparate parishes and synods. One thing is for certain, the scheme did not represent a sudden "big bang" development.

Table 8.1 Timeline of key events: 1718 to 1748

Date	*Event*	*Details*	*Source*
14 May 1718	Synod-based (i.e., local) funds "recommended" for widows, per proposal by P. Couper in 1716.	One-time contribution of of a tenth of stipend to create "stock" with interest (only) allocated to widow	Assembly Acts W. Wishart, elder, Mod.
12 May 1720	Act Concerning the book *Marrow of Modern Divinity*	"*Ministers are required to warn and exhort their people not to read or use*"	Assembly Acts
9 May 1723	Act Enlarging the Time Allowed for Maintaining…	Ministers given until Whitsunday 1725 to "*pay the decimas of their stipend*"	Assembly Acts
4 May 1727	Act Concerning John Simson (ongoing for three years)	Accused of adopting Arian views, "*presbytery empowered to suspend him*"	Assembly Acts
3 May 1733	Act Concerning Ministers & Concerning Ebenezer Erskine	Beginning of the secession of the "popular party" (from 1733/1740)	Assembly Acts
3 May 1733	Listed in Abridged Index, reference to commission	"*Power to receive proposals, concerning a fund to maintain indigent widows*"	Assembly Acts
8 May 1735	Act Reviving Recommendation, for (local) Fund in Each Synod	"*No prospect of any success in said scheme for a general fund…*"	Assembly Acts
Sep. 1736	Captain Porteus, lynched, Act of 10 Geo. II; to surrender	Robert Wallace refuses to read Act of Parliament in church	Alex Carlyle Autobiography
8 May 1740	Act and Sentence of Deposition against Seceders…the church	Finally brings the affair to a close	Assembly Acts
? 1741	Alexander Webster (Age 34) Designs the Scheme	"…*I drew the great lines of a plan in the year 1741…*"	Letter to Price in 1771
Feb. 1742	Resignation of Robert Walpole	Ilay replaced by Marquess of Tweeddale	ODNB
6 May 1742	Overture and Scheme	Flat £20 Benefit per £4 contribution, plus extra tax if needed. **Multiple = 5.**	Morren (1838) Assembly Acts
12 Nov. 1742	Rectified Scheme	Four contribution classes, with benefits ranging from £10 to £25. **Multiple = 3.8**	Printer: Fleming & Alison

(continued)

Table 8.1 (continued)

Date	*Event*	*Details*	*Source*
3 Jan. 1743	R. Hamilton (moderator) and R. MacIntosh (clerk) sign-off	"*Committee appoint the rectified scheme to be transmitted to ministers...*"	Printer: Fleming & Alison
12 May 1743	Robert Wallace elected as moderator of General Assembly	The scheme is presented, reviewed, and approved, noted in the index	Assembly Acts R. Wallace
20 May 1743	Colin Maclaurin is consulted	Suggestion to add three universities	CH9/17/16
? Feb. 1743	R. Wallace and G. Wishart go to London seeking Act	The scheme seems to have 'changed' between Edinburgh and London	CH9/17/23
25 Mar. 1744	Act of Parliament	Incumbents can opt-out, but selections must be made within 12 months	Parliament
10 May 1744	Assembly Acts to keep records	"*separate registers, proper columns*"	Assembly Acts
9 May 1745	Concerning the Fund	"By delay in transmitting the lists...with utmost difficulty"	Assembly Acts
8 May 1746	Act and Recommendation for Keeping Registers of the Dead	Congratulations to Duke of Cumberland (Culloden, 16 April, 1746)	Assembly Acts
7 May 1747	Act Concerning Fund...to transmit regularly their lists	"*The fund sustains considerable loss...*" Reports the fund is at a deficit	Assembly Acts
? May 1748	Calculations with the Principles and Data	"*...If old men had not been so numerous...would have been in no danger...*"	Printer: Thomas Lumsiden & Co.
12 May 1748	Amendments Proposed	"*If he died before hath paid for support of the fund...a sum of three years deducted*"	Assembly Acts G. Wishart

Sources and Additional References

Primary

Church of Scotland. General, Assembly. (1743). *A Scheme, for Providing an Annuity to Ministers Widows, and a Stock to Their Children; Humbly Offered to the Consideration of the Ministers of the Established Church of Scotland, by a Committee of the Commission of the Late General Assembly,*

Appointed to Draw up the Same, and Transmit It to the Several Presbyteries. Printed by R. Fleming and A. Alison. Eighteenth Century Collections Online: https://link.gale.com/apps/doc/CB0126539754/ECCO?u=ed_itw&sid=bookmark-ECCO&xid=fcf599e7&pg=1. Accessed 12 July 2022.

Great Britain. (1744). *An Act for Raising and Establishing a Fund for a Provision for the Widows and Children of the Ministers of the Church of Scotland, and of the Heads, Principals, and Masters, of the Universities of Saint Andrew's, Glasgow, and Edinburgh*. Thomas Baskett and Robert Baskett. Eighteenth Century Collections Online: http://link.gale.com/apps/doc/CB0126931101/ECCO?u=ed_itw&sid=bookmark-ECCO&xid=5f48cfc9&pg=1. Accessed 12 July 2022.

Great Britain. (1749). *An Act for Explaining and Amending an Act Passed in the Seventeenth Year of His Present Majesty's Reign, Intituled, An Act for Raising and Establishing a Fund for a Provision for the Widows and Children of the Ministers of the Church of Scotland, and of the Heads, Principals, and Masters of the Universities of Saint Andrews, Glasgow, and Edinburgh*. Thomas Baskett and Robert Baskett. Eighteenth Century Collections Online: https://link.gale.com/apps/doc/CB0126931184/ECCO?u=ed_itw&sid=bookmark-ECCO&xid=5fb0b1ea&pg=1. Accessed 18 July 2022.

Morren, N. (1838). *Annals of the General Assembly of the Church of Scotland, from the Final Secession in 1739, to the Origin of the Relief in 1752: With an Appendix of Biographical Sketches, Illustrative Documents, and Notes*. J. Johnstone.

National Records of Scotland (NRS) Church of Scotland (CH) Records

CH2/252/8: Lothian and Tweeddale synod, Minutes (1711–1721).

CH9/17: Miscellaneous Documents; NRS Catalogue—search results overview (nrscotland.gov.uk).

- CH9/17/1: Representation and Scheme for Providing an Annuity to Ministers Widows ... Offered to the Consideration of the ... Assembly of the Church of Scotland to Meet at Edinburgh the 12th May 1743. R. Fleming & Co.
- CH9/17/2: Scheme for Providing an Annuity to Ministers Widows, and a Stock to Their Children; Passed and Resolved by the General Assembly ... 12 May 1743. R. Fleming & Co..

Scott, H. (1915). *Fasti Ecclesiae Scoticanae: The Succession of Ministers in the Church of Scotland from the Reformation*. Oliver and Boyd.

Secondary

Allen, A. (2022). Corporate Charity for "the House": Craft Pensions and the Widows' Fund, 1670–1782. *Scottish Historical Review, 101*, 210–247.

Buchan, J. (2003). *Capital of the Mind: How Edinburg Changed the World*. John Murray Publishers.
Carlyle, A. (1910). *The Autobiography of Dr. Alexander Carlyle of Inveresk, 1772–1805*.
Defoe, D. (1889). An Essay upon Projects (London, 1697). *Available from Early English Books Online*, 735–742.
Dunlop, A. I. (Ed.). (1992). *The Scottish Ministers Widows Fund: 1743–1993*. Saint Andrew Press.
Henderson, L. (2016). *Witchcraft and Folk Belief in the Age of Enlightenment Scotland, 1670–1740*. Palgrave MacMillan.
Kidd, C. (2016). The Fergusson Affair: Calvinism and Dissimulation in the Scottish Enlightenment. *Intellectual History Review, 26*(3), 339–354.
Macree, D. (1973). Daniel Defoe, the Church of Scotland, and the Union of 1707. *Eighteenth-Century Studies, 7*, 62–77.
McDiarmid, A. (2021). The Equivalent Societies of Edinburgh and London, the Formation of the Royal Bank of Scotland, and the Nature of the Scottish Financial Revolution. *Journal of British Studies, 60*(1), 88–114.
Raffe, A. (2012). *The Culture of Controversy: Religious Arguments in Scotland, 1660–1714*. Boydell Press.
Raleigh, S. T. (1921). *Annals of the Church in Scotland*. Oxford University Press.
Sefton, H. R. (1962). *The Early Development of Moderation in the Church of Scotland* (PhD diss.). University of Glasgow.
Sher, R. B. (2015). *Church and University in the Scottish Enlightenment* (2nd ed.). Edinburgh University Press.
Sher, R., & Murdoch, A. (1983). Patronage and Party in the Church of Scotland, 1750–1800. *Church, Politics and Society: Scotland*, 267–298.
Smith, N. (1973). Clerical Corridors of Power: Extracts from Letters Concerning Robert Wallace's Involvement in Ecclesiastical Politics, 1742–1743. *Notes and Queries, 218*, 214–219.
Van Doodewaard, W. E. (2009). *The Marrow Controversy and Seceder Tradition: Marrow Theology in the Associate Presbytery and Associate Synod Secession Churches of Scotland from 1733–1799* (PhD diss.). University of Aberdeen.
Watt, D. (2007). *The Price of Scotland: Darien, Union and the Wealth of Nations*. Luath Press.
Whitley, L. A. B. (2013). *A Great Grievance: Ecclesiastical Lay Patronage in Scotland until 1750*. WIPF and Stock Publishers.

CHAPTER 9

Scientific Models Versus Religious Beliefs

Introduction and Objectives

This chapter addresses the matter of *scientific* versus *religious* beliefs, and in particular whether—as alluded to by Harari in *A Brief History of* Humankind—the fund represented a triumph of the scientific method over religious superstition. In particular, I will argue that positioning the 1744 scheme as having succeeded because the managers adopted a scientific versus a religious approach is misguided. First, the implicit conflict assumed between those two viewpoints is incorrect. The literature on that mythological conflict is vast, and I'll review some of the literature. Second, the 1744 scheme is unlikely to have succeeded without it having been embedded in a church ecosystem. I'll begin with some apparent "prophecies."

The annual reports of the trustees—which were discussed at length in Chapter 6 and Chapter 7—usually included a table at the very end that compared the "value" of the fund in each fiscal year relative to the value originally calculated in 1748. Technically, they used the terms "stock according to the calculation" and "stock according to the real amount." For example, a table at the end of the 1785 report listed the calculated value as £77,943 versus the real amount of £83,758. This is £5815 more, or approximately 7% greater than the amount originally projected in 1748. So, the fund was at a surplus of approximately 7%, relative to the (hypothetical) number projected nearly 40 years earlier, although the term

M. A. Milevsky, *The Religious Roots of Longevity Risk Sharing*,
https://doi.org/10.1007/978-3-031-62403-2_9

surplus (or deficit) usually refers to the relationship between asset and liabilities, not assets versus *projected* assets. The same table also shows the one single year (1765) when the calculated value of the fund was £58,348 and the real value of the fund was £58,347, which is within one single pound. If one can generalize, during the first 20 years the fund's real value was lower than the projected value, and then after 1765 the real value exceeded the projected value, by growing amounts. See Fig. 9.1 for a visual representation of these two numbers and specifically the percentage difference between the two of them.[1] At the same time, it's noteworthy to recall the phrase and mantra on how affairs were managed referred to in Chapter 7: "Must rise to this amount."

In the initial years, the "real value" as the trustees labelled the equity of the fund, was lower than the calculated values, but by the mid-1760s,

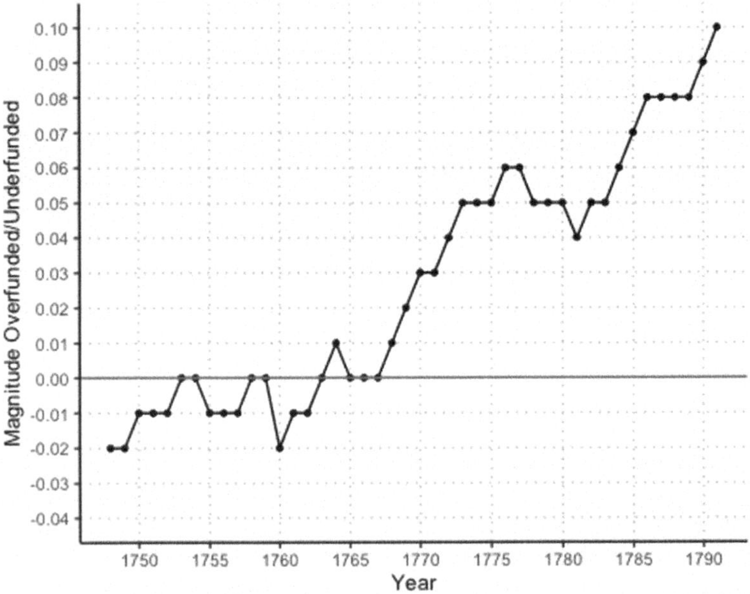

Fig. 9.1 Calculated projection vs. real value

[1] Figure created by Madeleine Salvadori, Research Assistant at the IFID Centre, based on data from CH9/16/2, p. 13. Used by author with permission.

the fund was higher, and in fact remained in that position well into the eighteenth century. So, while the late Reverend Dunlop in his 1992 book on the history of the fund was indeed correct that "actuality corresponded remarkably well," one must be careful not to oversimplify the context.[2] In particular, the claim made by both Harari, noted above, and by Ferguson in his book *Accent of Money* that Alexander Webster and Robert Wallace scientifically "predicted" (in 1748) the value of the fund to within one pound for 1765, was indeed true, but also very selective based on Fig. 9.1. Technically, it achieved that feat in the single year when the fund swung from a *less-than-projected* value to a *greater-than-projected* value. So, at best, their praise is cherry-picking and at worst it's deceiving. After all, any discrete sequence of numbers that transitions from negative to positive must traverse zero, also known as the *intermediate value theorem*.[3]

Again, as I have argued, it seems that Alexander Webster and the other trustees viewed the original calculation as a financial budget template. They "actively managed" the affairs of the fund, be it the expense management, investments, or even the interest rates, so it would achieve the calculated values. In fact, the trustees modified some of the scheme rules along the way to make that happen. For example, the scheme introduced a new tax for entrants if they were married when they joined, increased the interest rate on loans from 4 to 5%, and they obtained the authority to proportionally reduce benefits if needed—all so as to keep those two numbers as close as possible. Indeed, Webster took immense pride in how closely the two numbers tracked each other and he repeatedly referenced his own 1748 calculations to (essentially) compliment himself on his work.[4] Interestingly, as the fund exhibited larger positive gaps in later years, per Fig. 9.1 towards the right-hand side, pressure grew to increase payouts to widows and orphans because there was seemingly more available to spend. Yet Webster successfully resisted that pressure.[5]

[2] See Dunlop (1992), p. 17, and more generally his comments on the projection of the number of *widows* versus what had been forecast, as opposed to projections about the value of the fund itself.

[3] For a historical perspective on this theorem, see the article: Russ, S. B. (1980). A translation of Bolzano's paper on the intermediate value theorem. *Historia Mathematica*, 7(2), 156–185.

[4] See for example the last few pages of the annual reports for years from 1760 and 1780.

[5] Payouts to widows were increased after 1800. See Dunlop (1992) for the latter history.

What's pertinent is that scheme managers continued to report the 1748 "calculation value" and went so far as to anticipate these numbers for 100 years into the future. So, for example, the 1759 report listed a series of calculation values to the year 1831, which is breathtakingly audacious. In 1759 they anticipated the value of the fund for "Year of Our Lord 1800" to be £86,868, then £90,711 for the year 1810, £92,958 for 1820, and £93,900 for 1831. These anticipated values included shillings, pence, and even farthings and didn't include any range for uncertainty. They were projected values for 70 years into the future, which even a modern actuary might shy away from.[6]

To understand the uncertainty embedded within such a fund, note that starting with a group of 1000 (ministers), if 3% of that group is expected to die in the next year, a 97.5% confidence interval gives any number between 18 and 42. The number of survivors at the end of the first year would then be in the wide range from 958 to 988. So, at the very best these numbers should be interpreted as midpoints. The trustees appeared to place undue faith in the expectation that any deviations from the average would be immaterial. The economic psychologists Amos Tversky and Daniel Kahneman label this a fallacy of small numbers.[7] The fallacy manifests itself when consumers are asked to provide a statistical range for what might happen and their estimates are generally too close to the average: when they are asked to simulate 10 fair coin tosses, their results cluster around five heads and five tails.

The extreme confidence in anticipating the future poses the question, *Did the ministers believe they could prophesise or project values that far out in time?* This gets to the heart of probabilistic expectations in the mid-eighteenth century. This brief chapter argues that the answer to the question is a simple yes: it was viewed as divinely ordained. And, if need be, the ministers would make that future happen by reducing benefits, increasing contributions, or changing the legislation—but the calculated values were sacred and "the fund must rise."

[6] It's anachronistic to expect statistical confidence intervals, a concept developed in the early twentieth century, but there was no acknowledgement of any uncertainty or variance.

[7] Tversky and Kahneman (1971).

Stochastics in the Eighteenth Century

At that time mathematicians were just starting to appreciate the concept of randomness but had not quite developed a sense of risk. Ian Hacking, in his 2007 book *The Emergence of Probability*, writes that in the shadow of Isaac Newton's scientific revolution, "theological views of divine foreknowledge were being reinforced by the amazing success of mechanistic models."[8] To our main point, notice the word "reinforced" versus the opposite word, discarded or overshadowed or superseded.

On that note, it's instructive to review and engage with the views of the great statistician Karl Pearson (b. 1857, d. 1936), a Fellow of the Royal Society (in 1896). Pearson is described as the person "almost single-handily responsible for establishing the modern discipline of mathematical statistics" and associated with the statistics of correlation.[9] Today's standard undergraduate university course on basic statistics could easily be described as 60% Pearson and 40% all the others who have come since. In the 1920s, Pearson delivered a series of lectures, which were eventually published by his son Egon Pearson, under the title *The History of Statistics in the 17th and 18th Centuries*. This is precisely our period, and it is no surprise that annuity schemes such as the 1744 fund for widows and orphans, makes a number of appearances.[10]

More to the point of science versus religion, in one of those lectures,[11] Professor Pearson introduced his audience to the work of the German demographer and theologian Johann Peter Süssmilch (b. 1707, d. 1767). In 1741, Süssmilch, a Protestant pastor, wrote and published the first edition of a book that was appropriately titled in German *Die göttliche Ordnung* (English: Divine Design), large sections of which Pearson translated. Iam Hacking, quoted earlier, credits Süssmilch as the first

[8] Hacking (2007), *The Emergence of Probability: A Philosophical Study of Early Ideas about Probability Induction and Statistical Inference*, p. 3.

[9] For Karl Pearson, see entry by James Woiak in the *Oxford Dictionary of National Biography*, https://doi.org/10.1093/ref:odnb/35442.

[10] Karl Pearson's views on eugenics have not withstood the test of time, and the focus here (exclusively) is on his indisputable impact on the science of statistics.

[11] Pearson (1978), *The History of Statistics in the 17th and 18th Centuries against the Changing Backgrounds of Intellectual, Scientific and Religious Thought*.

substantial work on demography, a sentiment that is echoed to this day by German demographers.[12]

It seems that Pastor J.P. Süssmilch, according to Professor of Statistics Karl Pearson, was one of the earliest advocates of a notion described as the "stability of statistical ratios as evidence of divine provision." In other words, Süssmilch treated rates of birth, death, gender ratios, and mortality as part of a "beautiful and perfect order" that generated "reverence for the all wisest author of this order of nature."[13] Recall that this refers to the same science of probability theory that underlies longevity risk pooling, annuity fund management, etc. Quoting directly from Süssmilch, on the matter of predicting future mortality, *Die göttliche Ordnung* (translated by Pearson) says:

> Does not indeed the stability of the rule of mortality force on us the most pleasurable admiration? Does it not take us by the hand and lead us directly to the Creator of this stable law of Nature? Must one not reverence him who is the Lord of Life and Death? Can one continue to think indeed of death as due to uncontrolled and blind chance without being guilty of madness?[14]

From Pearson to Stigler to Hald

Anders Hald, in the *History of Probability and Statistics*, notes that this admiration of the regularity in demographic statistics resulted in the neglect or even simple lack of interest in variance or dispersion. The *Law of Large Numbers* upon which the valuation of reversionary annuities is based wasn't formulated until the work of Jacob Bernoulli (b. 1655, d. 1705) was published posthumously in 1713, and it took time to diffuse into practice.[15] Stephen Stigler, in his *History of Statistics*, notes that Bernoulli was the "first attempt at the quantification of uncertainty" while

[12] Hacking (2006), p. 113. See also Hecht (1987), 'Johan Peter Süssmilch: A German Prophet in Foreign Countries'.

[13] Pearson (1978), p. 317.

[14] Ibid., p. 318.

[15] See Sheynin (1968).

the second attempt was the work of Abraham de Moivre, published in multiple editions starting in the 1730s.[16]

In the mid-eighteenth century, the concept of variance had yet to be discovered, formulated, measured, or tamed. In a less theistic interpretation, the statistician Harald Westergaard, writing in his 1932 book *Contributions to the History of Statistics*, noted that "our predecessors first of all were struck by the regularity and cared less for the deviation."[17]

Scottish physician and satirist Dr. John Arbuthnot (b. 1667, d. 1735) is another relevant example of the religious-infused view of probability and statistics.[18] In the *Transaction of the Royal Society* (TRS) in 1710, he associated the statistical regularity of demographic data and the ratio of baby boys to girls (slightly more boys) with divine providence. He titled his article, in a so-called scientific journal, "An Argument for Divine Providence."[19] Pearson summarizes this typical mid-eighteenth-century view of modelling randomness:

> I am aware that some of my audience have thought I was exaggerating the stress laid by these early statisticians on the stability of statistical ratios as evidence of divine provision. But if they read Süssmilch's they will revise their opinions.[20]

Lorraine Daston, in her *Classical Probability in the Enlightenment*, notes that Edmond Halley believed more in the regularity of mortality than his own irregular data could support.[21] Daston echoes the above-noted eighteenth-century ignorance of variance, but doesn't go as far as Pearson in attributing the stability to God willing it. She is justifiably puzzled by why mortality rates would have been assumed to be regular, stable, and predictable, while other phenomena such as fires and plagues were not, but attributes this to a selective tendency to neglect deviations in empirical data in favour of expected regularities that had attracted researchers to the original data—but not, as Pearson argues, religious

[16] Stigler (1986), p. 77.

[17] Westergaard (1932), p. 72.

[18] Arbuthnot (1710).

[19] Arbuthnot (1710).

[20] Pearson (1978), p. 317.

[21] Halley (1693).

beliefs. I humbly differ with Daston, especially considering the scheme architects were all ordained ministers.

Variance Ignorance and Minister Price

Not everyone fell into this variance ignorance trap, however. Daston notes that the Dutch mathematician Nicholas Struyck (b. 1687, d. 1769) complained in 1740 that "mortality doesn't listen to our suppositions" and that many of the tables allegedly based on observation were in fact "pure hypotheses." In the end, it seems that Daston reluctantly concludes that it was a combination of natural theology and what she calls proto-quantification that may have contributed to this misplaced confidence in the regularity of selected statistical phenomena such as mortality and longevity patterns.[22]

The dissenting Unitarian minister Richard Price,[23] corresponded with Alexander Webster about the pricing of reversionary annuities.[24] In particular, Price criticized some of the calculations underlying the 1744 Presbyterian scheme, in the early edition of his (famous) textbook. Of Minister Price, Professor Pearson said in his lectures: "Price's book on reversionary payments is perhaps the most remarkable textbook ever issued on actuarial science."[25] Price noted that in widow schemes, the first entrants always profit at the expense of those dying later, which then creates a risk that contributions will not be enough for the annual payments. In the first edition of his book in 1771, Price named and criticized the architects of the Church of Scotland's scheme, and Webster replied to defend his assumptions.[26] In brief, their disagreement came down to assumptions about whether there might be 380 or as many as 400 widows, based on best-available mortality tables, and eventually Minister Price acquiesced and he came around to Webster's views. The advanced editions of Price's work removed the critical passage and even

[22] Daston (1988), chap. 3.

[23] For Richard Price, see entry by Richard Thomas in the *Oxford Dictionary of National Biography*, https://doi.org/10.1093/ref:odnb/22761. For his impact on the American Constitution, start with Cone (1948), 'Richard Price and the Constitution of the United States'.

[24] Price (1771).

[25] Pearson (1978), p. 395.

[26] See Webster (1771), *A Letter from the Rev. Dr Webster, of Edinburgh* ....

praised Webster. This is recounted in more detail, in Dunlop (1992), and Price's (eventual) "endorsement" of Alexander Webster and the fund was viewed as an important milestone in the life of the scheme.

In that same monograph by Minister Richard Price, written a century before Bismarck's pensions—or a decade before Condorcet's proposals—Price offered the first scheme for old-age pensions, and as a result, he is justly called the father of old-age pensions. As noted elsewhere, the link from Price to Webster and the 1744 scheme is critical. Minister Price's pension credit has been echoed by financial economists in the twenty-first century, not just by statisticians in the early twentieth century. Geoffrey Poitras, in *The Early History of Financial Economics*, calls Price's parish-based plan "exceptional and ingenious" and laments that the House of Lords rejected the scheme.[27]

The key and central point here is that even Richard Price, who was critical of some aspects of how the 1744 scheme was designed, built his work on a "Newtonian doctrine of the activating deity," according to Professor Pearson: "We seem fated in our history of statistics to mingle with the theologians, orthodox and unorthodox, Anglican and nonconformist."[28] In another quote from Ian Hacking, to this exact point, "Newton's idea of an omnipresent deity who maintains mean statistical values, formed the foundation of statistical development," and as a result, post-Newtonian mathematicians were "more influenced by Newton's theology than by his mathematics."[29] It's worth recalling that Colin Maclaurin, the third member of the triumvirate who died as a result of wounds incurred during the Jacobite rebellion, reviewed and improved the calculations made by Webster and Wallace. This same Maclaurin was a hand-picked protégé and leading disciple of Isaac Newton.[30] The point here is that positioning the 1744 scheme as a break from religious superstition, in favour of the scientific methods, is a gross simplification.

To summarize, perhaps it's now easier to understand the audacity with which the Presbyterian ministers made century-ahead financial predictions for the fund. If mortality patterns exhibit a regularity that was divinely

[27] Poitras (2000), p. 470.

[28] Pearson (1978), p. 349. See also Hald (2003).

[29] Hacking (2007), p. 172, quoting a separate article on Abraham de Moivre, written by Karl Pearson in 1926.

[30] See Bruneau (2020).

ordained, stable, and predictable, then fund values could be determined with the same confidence. Thus, while Webster's calculations are devoid of the biblical language that peppered his popular sermons—a favourite of his was the miracle behind Haman's drawing a lot that gave Esther a full year to prepare—his mortality calculations might have been driven by a similar belief in divine force.[31]

Conclusion

This chapter (and my own thinking on this matter) is inspired by the recent writings of Peter Harrison, Charles Taylor, Ronald Numbers, and Rodney Stark.[32] All four of them are eminent scholars, albeit in disparate fields, who have attempted to reframe the discourse on the relationship between science and religion. Traditionally, this relationship has been viewed through the lens of the "conflict thesis," a perspective that posits an inherent and perpetual struggle between scientific progress and religious faith. This thesis, which gained prominence and popularized during the nineteenth century by John William Draper and Andrew Dickson White, has been challenged by these four (as well as many other) scholars. They argue for a more nuanced and historically informed understanding of the relationship between these two *magisteria*.

For example, Peter Harrison in his various writings, has argued that theological concepts and religious institutions played a pivotal role in shaping the early scientific method, challenging the notion that religion has always been an obstacle to scientific inquiry. This dovetails nicely with the arguments in this chapter. Similarly, Rodney Stark has demonstrated how religious beliefs have, in many instances, encouraged scientific exploration and innovation. Charles Taylor encourages his audience to respect the contributions and limitations of both realms, and Ronald L. Numbers provides numerous instances where religious institutions have been integral to scientific advancements. Together, they all advocate for a departure from reductionist narratives—such as the one offered by Harari—that exclusively highlight discord and opposition.

[31] Webster (1740).

[32] In particular, see Stark (2004), Taylor (2007) and the edited collections, cited as Harrison (2010), and Numbers (2009). For a recent, popular, exposition see Spencer (2023).

More specifically, this chapter speaks to a more nuanced interpretation of the origins of the widows' scheme, relative to what has been offered in the previous literature as well as the popular narrative. Indeed, popular authors[33] give insufficient credit to religious beliefs—as well as the local synodical origins discussed in the prior chapter—and provide scarce acknowledgement of the immense administrative undertaking involved in successfully managing such a fund.

The intellectual genesis of the fund took place a quarter of a century prior to a scientific big bang in 1744. Moreover, initially it was actuarially unsustainable, designed with a belief in divine providence, and most likely owes its longevity to an "adaptive church bureaucracy", or to some extent, belief in a divine creator who would ensure laws of probability would transform to natural realities. In the words of Euegene M. Klaaren, who wrote *Religious Origins of Modern Science*:

> Partly due to convictions of divinely imposed laws of nature, knowledge in the new science was marked by a disciplined search for and positive statement of laws…But the law of the Creator as Author was consistently taken for granted in spheres as diverse as nature, morals and religion.[34]

Sources and Additional References

Arbuthnot, J. (1710). An Argument for Divine Providence, Taken from the Constant Regularity Observ'd in the Births of Both Sexes. *Philosophical Transactions of the Royal Society of London, 27*(328), 186–190.

Bruneau, O. (2020). Colin Maclaurin (1698–1746): A Newtonian between Theory and Practice. *British Journal for the History of Mathematics, 35*(1), 52–62.

Daston, L. (1988). *Classical Probability in the Enlightenment*. Princeton University Press.

Dunlop, A. I. (Ed.). (1992). *The Scottish Ministers Widows Fund: 1743–1993*. Saint Andrew Press.

Hacking, I. (2006). *The Emergence of Probability: A Philosophical Study of Early Ideas About Probability, Induction and Statistical Inference* (2nd ed.). Cambridge University Press.

[33] Further propagated by blogs such as the Legacy Trust and The Tontine Coffee House.

[34] Source: Klaaren (1977), pp. 189–191.

Hald, A. (2003). *A History of Statistics and Probability and Their Application before 1750*. John Wiley & Sons.

Halley, E. (1693). An Estimate of the Degrees of the Mortality of Mankind; Drawn from Curious Tables of the Births and Funerals at the City of Breslaw; with an Attempt to Ascertain the Price of Annuities Upon Lives. *Philosophical Transactions of the Royal Society of London, 17*(196), 596–610.

Harrison, P. (Ed.). (2010). *The Cambridge Companion to Science and Religion*. Cambridge University Press.

Hecht, J. (1987). Johann Peter Süssmilch: A German Prophet in Foreign Countries. *Population Studies, 41*(1), 31–58.

Klaaren, E. M. (1977). *Religious Origins of Modern Science: Belief in Creation in Seventeenth-century Thought*. Eeerdmans Publishing Company.

Numbers, R. L. (Ed.). (2009). *Galileo Goes to Jail and Other Myths About Science and Religion*. Harvard University Press.

Pearson, K. (1978). *The History of Statistics in the 17th and 18th Centuries against the Changing Backgrounds of Intellectual, Scientific and Religious Thought*. Edited by Egon Pearson. Charles Griffin & Company.

Poitras, G. (2000). *The Early History of Financial Economics, 1478–1776: From Commercial Arithmetic to Life Annuities and Joint Stock*. Edward Elgar.

Price, R. (1771). *Observations on Reversionary Payments; on Schemes for Providing Annuities for Widows, and for Persons in Old Age; on The Method of Calculating the Values of Assurances on Lives; and on The National Debt. To Which Are Added, Four Essays on Different Subjects in the Doctrine of Life-Annuities and Political Arithmetick*. 2nd ed. Printed for T. Cadell, in the Strand. Eighteenth Century Collections Online. http://link.gale.com/apps/doc/CW0105622537/ECCO?u=ed_itw&sid=bookmark-ECCO&xid=2190f221&pg=1. Accessed 12 July 2022.

Russ, S. B. (1980). A Translation of Bolzano's Paper on the Intermediate Value Theorem. *Historia Mathematica, 7*(2), 156–185.

Sheynin, O. B. (1968). Studies in the History of Probability and Statistics: On the Early History of the Law of Large Numbers. *Biometrika, 55*(3), 459–467.

Spencer, N. (2023). *Magisteria: The Entangled Histories of Science & Religion*. Simon and Schuster.

Stark, R. (2004). *For the Glory of God: How Monotheism Led to Reformations, Science, Witch-hunts, and the End of Slavery*. Princeton University Press.

Stigler, S. M. (1986). *The History of Statistics: The Measurement of Uncertainty before 1900*. Harvard University Press.

Taylor, C. (2007). *A Secular Age*. Harvard University Press.

Tversky, A., & Kahneman, D. (1971). Belief in the Law of Small Numbers. *Psychological Bulletin, 76*(2), 105–110.

Webster, A. (1740). *The Wicked Life, and Fatal but Deserved Death of Haman, Ahasuerus's Prime Minister. A Sermon, Preached Before the Right Honourable,*

Lord Provost, Magistrates and Town-Council of Edinburgh, at Their Election, on Tuesday, 30th September, 1740. Eighteenth Century Collections Online: https://link.gale.com/apps/doc/CB0127354332/ECCO?u=ed_itw&sid=bookmark-ECCO&xid=4eceb61b&pg=1. Accessed 12 July 2022.

Webster, A. (1771). *A Letter from the Rev. Dr Webster, of Edinburgh, to the Rev. Dr Price, of London, and Dr Price's Answer, Relative to the Establishment for a Provision to the Widows and Children of the Ministers and Professors in Scotland.* Printed by Murray & Cochran. Eighteenth Century Collections Online: http://link.gale.com/apps/doc/CW0120080822/ECCO?u=ed_itw&sid=bookmark-ECCO&xid=98c60af8&pg=1. Accessed 12 July 2022.

Westergaard, H. (1932). *Contributions to the History of Statistics.* P.S. King & Son Ltd.

CHAPTER 10

The First Biblical Annuity

The historian of actuarial science, C.G. Lewin, in his (2003) encyclopedic book entitled *Pensions and Insurance Before 1800*, writes, "The idea of a pension for life is very old indeed." It is certainly much older than the year AD 1744, when the Presbyterian annuity fund was launched to offer them to widows. In particular, he labels a 2600-year-old pension annuity that is mentioned in the Hebrew Bible and was granted to the King of Judah in the mid-sixth century BC, as one of the earliest documented life annuities in ancient history. In the same chapter, Lewin discusses pension annuities that were granted (or taken away) from the Greek speech writer Lysias in 400 BC, as well as those granted to Roman military veterans, the merits of which were debated by the great Roman statesman Cicero in the first century BC. After all, the word annuity comes from the Latin word *annuitas*, which means annua or yearly payment, also related to *annus*, the Latin word for year. The important thing to note here is that the granting of pension annuities for life is as ancient as life itself.[1]

Nevertheless, as noted by Lewin, if one searches for the very first historically documented life annuity in which the financial terms and conditions can be reasonably ascertained, it's the biblical pension annuity

[1] Lewin (2003, Chapter 1, and in particular pp. 6–7). Note that Kopf (1927, p. 231), makes reference to an earlier Egyptian (1100 BC) annuity purchased by a certain Prince Hepdefal, who ruled in Sint in the Middle Empire, but provides no additional information or sources for the terms of the Egyptian annuity, or if it was for life.

M. A. Milevsky, *The Religious Roots of Longevity Risk Sharing*,
https://doi.org/10.1007/978-3-031-62403-2_10

granted to the King of Judah. In fact, many other authors—myself included[2]—have referenced this 2600-year-old annuity as proof of its ancient pedigree, although usually in a very cursory and ahistorical manner. In this chapter I'll delve deeper and attempt to answer the following questions. Who exactly was this annuitant king? Why was the annuity granted to him? More importantly, why is this pension annuity important enough to be emphasized (twice) in the Hebrew Bible?

So, in this chapter I'll return to a time before longevity risk sharing and pooling was successfully inaugurated in 1744. This was a time when pension annuities were issued solo to individuals and not groups, longevity risk was therefore not diversified, and the issuer's default risk loomed large on the horizon. For that journey back in time, I'll delve into the background, context, and even attempt a (soft) actuarial valuation of the King of Judah's pension annuity—again the first documented personal longevity hedge in ancient history.

And, while it would be natural for some readers to question the historicity or accuracy of a biblical tale, there is compelling archeological evidence to support the premise that this "story" actually took place and the annuitant in question was indeed the beneficiary of lifetime payments. Moreover, in yet another link or connection with the 1744 Presbyterian annuity fund at the core of this book, it's quite likely our protagonist and noted Reverend Alexander Webster was well aware of this biblical (annuity) story. How do I know this? Well, the only extant and available copies of his (famous) Sunday sermons were all narrowly focused on the biblical Book of Esther. That book or scroll depicts events that took place during the Persian Achaemenid Dynasty, between 559 and 331 BC. And, coincidentally, the story of Esther takes place a mere few decades after the above-noted annuity was granted to the King of Judah. So, for the sake of Alexander Webster's memory, this is yet another reason to delve into the history of the first documented biblical annuity around the sixth century BC.[3]

The remainder of the story told in this chapter is organized as follows. In the next section I'll present and interpret the assorted biblical texts in

[2] See also Milevsky (2013, pp. 6–8) for an overview of the 'history' of annuities during the last 4000 years, as well as the review by Gillman (2014).

[3] See the sermons referenced as Webster (1740) and Webster (1746), both focused on the Book of Esther.

which this unique annuity is referenced. I'll also explain why this particular King of Judah is so significant for ancient (Jewish) history, and why this annuity—much more than a series of financial payment—symbolized the rehabilitation of a monarch and the redemption of a nation in exile. The section after that will present the archeological evidence to backup the existence or at least the plausible existence of this sixth-century pension annuity. Then, I will discuss how one might value the King of Judah's annuity, by leveraging data on mortality, interest rates, and even the price of oil—yes, to be explained—in the sixth century BC. Finally, the last section will wrap up the chapter with a link to yet another classical book, *The Richest Man in Babylon*, and offer some concluding remarks about retirement security throughout history.

Jehoiachin's Pension Annuity

The story behind the King of Judah's annuity, whose name for the record was Jehoiachin (a.k.a. Jeconiah, a.k.a. Coniah), is mentioned in two different places within the Hebrew Bible. The first reference, and the one cited by Lewin appears in the Book of Kings II, Chapter 25, Verses 27–30. Here is the entire text from the New King James Version (NKJV):

> [27] Now it came to pass in the thirty-seventh year of the captivity of Jehoiachin king of Judah, in the twelfth month, on the twenty-seventh day of the month, that Evilmerodach king of Babylon, in the year that he began to reign, released Jehoiachin king of Judah from prison. [28] He spoke kindly to him and gave him a more prominent seat than those of the kings who were with him in Babylon. [29] So Jehoiachin changed from his prison garments, and he ate bread regularly before the king all the days of his life. [30] And as for his provisions, there was a regular ration given him by the king, a portion for each day, all the days of his life.

I have underlined the last few words in the paragraph for emphasis, and this term and verse also appear in the epigraph to this book, and I actually considered naming the book *Daily Bread for Life*, instead of the rather cumbersome one it currently has.[4] Regardless of the significance of the term "bread for life," Jehoiachin's annuity, which is how I'll reference

[4] My original title was kiboshed by the editor, who felt it might "confuse" search engines.

it from here on, is also described in the Book of Jeremiah, Chapter 52, Verses 31–34. There it appears with two slight variations, according to the text from NKJV:

> [31] And it came to pass in the seven and thirtieth year of the captivity of Jehoiachin king of Judah, in the twelfth month, in the five and twentieth day of the month, that Evilmerodach king of Babylon in the first year of his reign lifted up the head of Jehoiachin king of Judah, and brought him forth out of prison, [32] And spake kindly unto him, and set his throne above the throne of the kings that were with him in Babylon, [33] And changed his prison garments: and he did continually eat bread before him all the days of his life. [34] And for his diet, there was a continual diet given him of the king of Babylon, every day a portion until the day of his death, all the days of his life.

Without getting caught up in the minutia of biblical translations—since the original was written in Old Hebrew—I should note the NKJV translation can differ slightly from the English Standard Version (ESV), so one should not read too much into individual words, such as rations versus diets versus provisions. Nonetheless, there are two noticeable differences between the text in Kings II and Jeremiah. The Kings II version states the date of the King's release from prison as the 27th of the 12th month, which would be the Babylonian month of Addaru. The Jeremiah text claims this critical event took place on the 25th of the 12th month. Now, this two-day gap isn't of much concern (at least to Jehoiachin) because the 37 years he spent in prison would amount to over 13,000 days.[5] Moving on, the second noticeable difference between the texts is that the Jeremiah narrative emphasizes and adds the words "until the day of his death" in addition to "all the days of his life." Read in totality, though, this appears to indicate the grant was likely an annuity contingent on life. Much more importantly, both of these paragraphs or texts appear in the absolute last chapter and last verse of the (very long) biblical Book of Kings II and Book of Jeremiah, respectively. The narrator (of Kings II) and the famous prophet end their entire historical narrative with what is an uplifting story of redemption; a Judean king who was released from

[5] Some biblical exegetes (e.g., Rashi) have explained the change (27 versus 25) as being the difference between when Evilmerodach ascended to the Babylonian throne, and when Jehoiachin was released; quite plausible.

prison after years in captivity. So, who was this king exactly and why was his release so important and inspiring?

The Last Living King of Judah

Jehoiachin wasn't just another king who happened to find himself on the wrong side of a battle with the Babylonians, he was in fact the last remaining living king of Judah, the southern province of what is today known as the State of Israel. His release from prison might be compared with the release of the South African Nelson Mandela, the Russian Anatoly Sharansky, or the Kenyan Jomo Kenyatta. Emphasizing the importance of this event, Fig. 10.1 is one of the many images depicting King Jehoiachin's release from Babylonian prison. At the very top of the stairs—standing next to Jehoiachin—is the Babylonian king Evilmerodach (a.k.a. Amel-Marduk), who appears to be showing him great respect. He is elevated above the other regal-looking figures, that is, the other kings who are with him.

The (odd) name of the Babylonian king Evilmerodach, who is depicted standing next to his throne, has unfortunately been misinterpreted (by the English) as denoting wickedness or sinfulness. In fact, unlike his father Nebuchadnezzar—a notorious character who I'll circle back to in a few pages and who was indeed evil from a Judean perspective—his son King Evilmerodach didn't last very long on the throne. Nevertheless, the few times I do refer to him, when discussing the pension annuity he granted Jehoiachin, I'll use his (more popular among historians) name Amel-Marduk, who reigned from 561 to 560 BC. The release of King Jehoiachin from prison in the year 561 BC marked the beginning of the end of the Babylonian exile.

A full 37 years earlier, King Jehoaiachin had only spent three months on the throne of Judah when he was deposed and carried away by the conquering king Nebuchadnezzar II in March of 597 BC. He was (at most) 18 years old when he was dragged off to Babylonian prison. And, according to the historian Josephus, Jehoiachin, he was "kind, genuine, gentle, benevolent, worthy, decent, honest and friendly," which are descriptions and characteristics one might expect from a mythologized and hagiographic description of the last king of Judah.[6] His deportation

[6] The Talmudic literature further describe him as a "bad king" like his father Jehoiakin who was also a "bad king," unlike his grandfather Josiah, who was a "good king."

THE RELEASE OF JEHOIACHIN.—II. Kings xxv. 28.

Fig. 10.1 Release of King Jehoiachin and his annuity

in March 597 BC, took place slightly more than a decade before the final sack and destruction of Jerusalem in August of 586 BC. From the time of Jehoiachin's imprisonment, Judea became effectively a vassal state of the Babylonian empire. The Babylonian king Nebuchadnezzar (who was Amel-Marduk's father) appointed Jehoiachin's uncle as king in his place, while Jehoiachin sat in a Babylonian prison. Alas, the vassal king (uncle Zedekiah) made the unfortunate error of rebelling and siding with the Egyptians against the Babylonians. And, a decade later, early August 586 BC to be precise, Nebuchadnezzar and his army and generals returned yet again to Judah, and this time he utterly destroyed Jerusalem and Solomon's temple, for good.[7]

So, here is the key symbolic event being narrated in those two biblical verses. The son of Nebuchadnezzar II (the destroyer), that is the new Babylonian king Amel-Marduk, rehabilitated King Jehoiachin. A son takes over for his father and immediately sets out to reverse his policies and mend bridges. Delving into the political history of this period might help shed light on why Jehoiachin was released from prison and/or why he was granted a pension and given such prominence and respect at the new king's table. Although, it's unclear how long that period (or even the annuity) lasted, as Amel-Marduk was dead a year later, then Nerigliassar took over for four years until 556, then Labashi-Marduk for less than a year, and finally Nabonidus (the famous last king of Babylon) from 555 to 539 BC.[8]

Judean Genealogy

Without getting lost in the minutia of biblical history during the late seventh and early sixth century BC, five kings reigned in the half-century before the destruction of the temple and final sacking of Jerusalem in August of 586 BC. These five kings were from the so-called Davidic line of decent and were related to each other as described in Table 10.1.

In Table 10.1, the number in square brackets after the letter K denotes the order in which the kings reigned. Back to Jehoiachin, our protagonist and annuitant, he ascended to the throne after his father Jehoiakim died (in battle against Babylon), but was very soon deposed (by the King of Babylon) and then imprisoned (in Babylon) and finally his uncle Zedekiah

[7] For more on Jehoiachin and Nebuchadnezzar, see Feldman (1995), Sack (2004), Clements (2007), and Alstola (2019). For the exact chronology of events, see Thiele (1956) and Finegan (2015). The Rabbinic chronology, per *Seder Olam*, assumes the destruction of the first temple took place in 421 BC, which implies that King Jehoiachin lived and died approximately 165 years later, sometime during the Persian empire. This chapter isn't the place to delve into this "problem".

[8] There is an odd story told by St. Jerome, in his commentary on the biblical book of Isaiah, the Amel-Marduk was in fact imprisoned by his father Nebuchadnezzar, together with Jehoiachin, and they became friends. Perhaps explaining the annuity grant upon Nebuchadnezzar's death and Amel-Marduk's accession to the Babylonian throne. See Sack (2004) for more details. The chronology of the Babylonian kings is from the history book by Paul-Alain Beaulieu (2018). See also Weiershäuser and Novotny (2020).

Table 10.1 Lifespan and the final kings of Judah[9]

Generation 1	**[K1] King Josiah** *Reigns 31 years. Dies in July 609 BC*		
Generation 2 (Brothers)	**[K2] King Jehoahaz** *Reigns 3 months* *Imprisoned 609 BC & Dies in Egypt*	**[K3] King Jehoiakim** *Reigns 11 Years* *Dies in Dec. 598 BC*	**[K5] King Zedekiah** *Reigns 11 years* *Flees in Aug. 586 BC*
Generation 3	**[K4] King Jehoiachin** *Reigns 3 months + 10 days* *Deposed 16 March 597 BC, Freed in 561 BC*		
Generation 4	Shealtiel & Pedaiah		
	Fall of Nabonidus and Neo-Babylonian Dynasty in 539 BC *Rise of Cyrus II and the Achaemenid (Persian) Dynasty*		
Generation 5	**Zerubbabel** (Governor of Judah)		

took over for the last 11 years of the pre-exilic Judean kingdom. That was the end of the Judean kingdom, for at least 70 years, or perhaps longer depending on one's definition of Judean.

At the fear of getting lost in the chronological weeds, some of the dates in Table 10.1 are speculative, and specifically the death of Zedekiah is unknown. However, the date of 16 March 597 BC on which Jehoiachin was deposed and then imprisoned has been corroborated by archeological evidence. The so-called *Babylonian Chronicles*, which are a collection of real cuneiform tablets[10] stored in the British Museum and transcribed by D.J. Wiseman, allow us to precisely date the end of Jehoiachin's (very short) reign.[11] The key takeaway here, for our purposes, is that although King Zedekiah reigned after Jehoiachin, his nephew, the last remaining heir to the Judean monarchy was verifiably imprisoned Jehoiachin. And, he was alive 37 years later in the year 561 BC, when he was released from Babylonian prison by King Amel-Marduk. Up to this point, I have

[9] Every one of these five kings had (and is known with) an alternate name. Some consider Gedaliah, who was appointed after Zedekiah fled in 685 BC, as a monarch and therefore the proper last king of Judah. See Knauf and Guillaume (2016, p. 137).

[10] Unlike George S. Clason's fictitious ones in *The Richest Man in Babylon*.

[11] See Thiele (1956).

provided archeological evidence that the biblical Jehoiachin was indeed king, captured and imprisoned in Babylon. But what about the annuity?

Archeological Evidence and Cuneiform Tablets

In the early years of the twentieth century, German excavations in Babylon—then under the control of the Ottoman Empire—by archaeologist Robert Koldewey, located the Ishtar Gate together with over 5000 cuneiform tablets from the state archives of King Nebuchadnezzar. Similar to the Assyrian archives unearthed at the capital city of Nineveh, they consisted of administrative documents recording the delivery of commodities to be stored and processed in Babylon, as well as their eventual distribution to people of Babylonian and foreign origin. And, although only 6% of these Babylonian tablets have been deciphered (thus offering Ph.D. topics for many future Assyriologists), there is evidence supporting the idea that King Jehoiachin was already receiving rations before he was given amnesty and released from prison. The rations were provided in, yes, oil.[12]

Figure 12.2 is an Akkadian cuneiform inscription list of rations, and mentions the name of Jeconiah (Jehoiachin), King of Judah and his family. The tablet is stored in the Vorderasiatisches Museum in Berlin.

The historian Tero Alstola (2019) writes: "The presence of King Jehoiachin and five Judean princes on a ration list from 591 shows that they were held hostage while Zedekiah was still ruling Judah as a vassal king." He also notes: "The oil rations given to Jehoiachin were large, being six times bigger than the average ration," which had been offered to others and documented in the archive. And, since this amount would have been too much for his own personal needs, the rations were likely meant to be redistributed to his family members and other dependants. In addition to the ration list from 591 BC, that is six years after he was taken to Babylon by Nebuchadnezzar, the name Jehoiachin and King of Judah appear three additional times in ration lists from the palace archives. Some have used these generous rations to argue that Judean royalty were never actually incarcerated in a prison, in the modern sense of the term, and that their captivity was more figurative than literal or perhaps was more like house arrest. Regardless of the extent of these restrictions, additional

[12] In addition to Alstola (2019), which is the basis for most of the next few paragraphs, see Begg (1986), Clements (2007), Horn (1967), Feldman (1995), and Grellet (2018).

rations were allocated to his five sons, two of which (presumably) would have been Shealtiel and Pedaiah (mentioned in Table 12.1). Perhaps even Jehoiachin's grandson Zerubbabel, who according to the Hebrew Bible led the first group of exiles back to Judah and Jerusalem after the Persians ended the Babylonian empire, was also a recipient.

Referring again to Tero Alstola, from the University of Helsinki, and his examination of the tablets, he writes: "It is noteworthy that the [Kings II and Jeremiah] text explicitly refers to Jehoiachin's regular allowance, which immediately reminds the modern reader of the ration lists in the Palace Archive." Yet, to be very clear, while the word annuity is not mentioned in any of the cuneiform tablets that have been deciphered and translated to date, Jehoiachin's ample rations are likely to have been continued after he was released by the new Babylonian king, Amel-Marduk. More importantly, his life annuity was paid in food rations, that is, bread and oil. This raises the question: *Can Jehoiachin's annuity be valued?* Needless to say, the Babylonian King Amel-Marduk didn't set up a fund to actuarially reserve against the annuity, nor is there any indication that it was risk pooled with others. But as an actuarial exercise, knowing the price of a "sila of oil," it is possible to value the life annuity issued to this 55-year-old (in captivity 37 years + age 18 years = 55). Here is an attempt.

Valuation of the Babylonian Annuity

Colin Maclaurin, our valuation expert from Chapter 6 and Chapter 7, would have recognized the following mathematical expression—displayed in Fig. 10.2—as the single most important equation in the field of annuity valuation. The left-hand side (a.k.a. the value of King Jehoiachin's pension annuity) is equal to a function of three variables, two of which have by now been determined. The letter C denotes the periodic (daily, monthly, or yearly) cash flows, which in our case would represent the market price of the provisions (a.k.a. daily bread for life) or his 10 "sila of oil," according to the tablets. The letter (p_i) in the numerator of the fraction, denotes the conditional survival probability, for a 54-year-old to various ages, and the letter R denotes the interest or valuation rate.

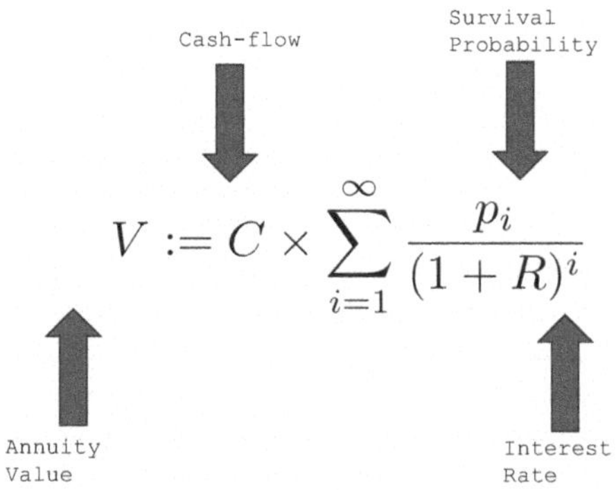

$$V := C \times \sum_{i=1}^{\infty} \frac{p_i}{(1+R)^i}$$

Fig. 10.2 Valuation picture for guaranteed longevity risk pooling

Annuitant Mortality: Borrow from a Roman

Indeed, the key ingredient to valuing any life annuity is the assumed mortality rate, or more generally, the life expectancy of the population. While Alexander Webster, Robert Wallace, and Colin Maclaurin had access to mortality tables—such as those composed by Edmond Halley—it would be ridiculous to think anyone in the sixth century BC might have considered how long King Jehoiachin was likely to live, after he was released. What mortality assumption should we use? In this section I'll take the opportunity to discuss the earliest attempt to arrive at a life expectancy (or mortality assumptions) for the purposes of annuity valuation. And that feat can be located over 1600 years ago, although perhaps 1000 years too late for Amel-Marduk and Jehoiachin. I'm referring to the height of the Roman Empire, and in particular to the jurist Ulpian, as described by Aemilius Macer, who was also a Roman jurist active during the third century AD, in the rein of Emperor Alexander Severus (AD 222–235). His prolific contributions to Roman legal texts were extensive, but he also provided some actuarial guidance. Here is a direct quote from the relevant text:

> Ulpianus says that this is the formula to be adopted in making the estimate of sustenance's to be furnished. The amount bequeathed to anyone for this purpose from the first to the twentieth year is computed to last for thirty years, and the Falcidian portion of that sum shall be reserved. From twenty to twenty-five years, the amount is calculated for twenty-eight years. From twenty-five to thirty years, the amount is calculated for twenty-five years;

from thirty to thirty-five years, the amount is calculated for twenty-two years; from thirty-five years to forty years, it is computed for twenty years; from forty to fifty years, the computation is made for as many years as a person lacks of the sixtieth year after having omitted one year; from the fiftieth to the fifty-fifth, the amount is calculated for nine years; from the fifty-fifth to the sixtieth year, it is calculated for seven years; and for any age above sixty, no matter what it may be, the computation is made for five years. Ulpianus also says that we use the same rule in making the calculation with reference to the legacy of an usufruct. Nevertheless, it is the practice for the computation to be made for thirty years from the first to the thirtieth, but after the age of thirty years it is made for as many years as the legatee lacks of being sixty; hence the computation is never made for a longer time than thirty years.[13]

Now, the above (long) paragraph might not appear to look like a table or matrix or vector of mortality rates, but with some tabulations and reorganization, it's possible to further "translate" the words into numbers and express the valuation rule suggested by said Ulpanius in Table 10.2, which echoes modern mortality tables. Unlike the exercise I conducted in the early parts of Chapter 7, this isn't a survival probability curve but more of a life expectancy "curve" or perhaps it is the actuarial annuity factors themselves. We really don't know how the Ulpanius table was actually used to convert a flow of Denarius into a proper present value.

Table 10.2 Ulpian's table circa 300 CE

Age range	*Annuity factor*
0–19	30
20–24	28
25–29	25
30–34	22
35–39	20
40–49	(60-x-1)
50–54	9
55–59	7
60+	5

[13] The original, of course, is written in Latin, and the above translation is from a chapter of a book, cited as de Vries and Zwalve (2004).

So, getting back to the Babylonian Empire (and away from the later Roman one) and with a large leap of time, which I have anachronistically "borrowed" from Ulpian's table, we can do the following. Assuming Jehioachin was indeed 55 years old when he was released from prison by king Amel-Markduk, his life expectancy would have been approximately seven years. The only item missing, then, is the interest or valuation rate (R) in the denominator of the above displayed equation. What was the appropriate valuation rate or interest rate to use in the sixth and fifth centuries BC? Ideally, we need the Babylonian term structure of interest rates.

Interest Rates in the Fifth Century BC

Sidney Homer and Richard Sylla in their classic (1963, first edition) book titled *A History of Interest Rates*, discuss interest rates going as far back as the third millennia BC, which would obviously cover the period in question. They suggest that in Sumer and Babylonia, our geographical location, during the sixth century BC, a normal interest rate was between 10% and 20% per year, and the legal maximum was 20%. That is obviously much higher than today, although one must be careful when defining an interest rate for thousands of years ago, in a time before national financial institutions and currencies. Notwithstanding the risk of anachronism, according to Homer and Sylla, by the fifth and fourth centuries and well into the Persian (Achaemenid) period, interest rates had increased to 40% in Sumer and Babylonia. So, while it seems these rates might not have changed much over 3000 years in Turkey (biblical Haran), the net present value (NPV) term on the right-hand side of Fig. 10.2 would be rather small.

For geographical and historical comparison purposes, during the same period in Egypt—if you could get your money down there during the mid-sixth century BC—loans for grain and silver were associated with interest rates as high as 85% to 160% per year. In April 2024, when this is being written, the Central Bank of Egypt was reporting an overnight lending rate of 27.25% per year on Egyptian pounds. So, they have come down a tad.[14] The anthropologist David Graeber, in his (2011) book titled *DEBT: The First 5,000 Years*, uses the term Mesopotamian debt

[14] Source: https://www.cbe.org.eg/en/.

crisis, which would have been especially acute in years of bad harvests, to describe this state of affairs during that period.

These very high interest rates also help us to understand the context behind the ancient laws of Hammurabi, a full thousand years earlier in 1750 BC. The Babylonian King Hammurabi was a ruler of the Babylonian Empire, and the stele with the Code of Hammurabi was discovered in the early twentieth century by a French (not German) archaeological team. The code, among many other stipulations and rules, placed limits on how much creditors could charge. All of this leaves us with a bit of a valuation conundrum. In some sense, one could really use any rate between zero and the proverbial 120, which isn't much help in valuing Jehoiachin's biblical annuity.[15]

And yet, oddly enough, there is one (and only one) reference to an actual interest rate in the Hebrew Bible, and it appears in the Book of Nehemiah. Even better, for our purposes, the historical period covered by Nehemiah is precisely the fifth century BC, the geographical location is Judah, and his connection to the annuity is more than coincidental. To make a very long story short, according to the Book of Nehemiah, the interest rate in Jerusalem in the aftermath of the return from the Babylonian exile was lower than the Egyptian rate (either in the present-day or the sixth century BC). The Judean interest rate was 12% per year. For those readers who want to know exactly where that number comes from—a hint is that interest rates per month were one in a hundred—I urge you to read the story of how Nehemiah urged creditors to ease up on the interest rates they were charging. For our purposes, though, the only thing you need from Nehemia is the 12% number I will use for the R value, remember Fig. 10.2, for putting a final number on Jehoiachin's pension annuity.[16]

Without getting too actuarial, using a life expectancy of nine years (from Ulpian's table) for King Jehoiachin, and a 12% interest rate for valuation purposes, the pure annuity factor—that is the number multiplying C in the equation within Fig. 10.2—is the number 4.50. In other words, the discounted value of the pension annuity at the time it was granted to King Jehoiachin, who I estimated to be 55 years of age when released

[15] Homer and Sylla (2005, 4th edition, pp. 54–61) and Graeber (2011, p. 81). For Hammurabi, see the Avalon Project translation at Yale University. https://avalon.law.yale.edu/ancient/hamframe.asp.

[16] See Neufeld (1954), who notes a monthly rate of 1% which is an APR of 12%.

and rehabilitated, was between four and five times the annual income (or amount of oil) he was receiving. So, if he was promised a sela of oil per week, or 52 per year, the value of the pension was $4.5 \times 52 = 234$ selas of oil. If he was granted a loaf of bread per day, or 365 per year, then the value of his pension annuity was $4.5 \times 365 = 1642.5$ loaves of bread, etc. Of course, this is a very crude approximate value—with many embedded assumptions—but the point is, it's a nice exercise. I should note once again, as I stated earlier in this book, while the mathematicians might be able to value the annuity, it wasn't until the late eighteenth century that anyone actually used these sorts of formulae to price and sell annuities to the public.[17]

The Richest Man in Babylon

Finally, for those readers who have heard of the book and are wondering about the connection to the financial planning classic, the link is rather interesting and not coincidental. For those who haven't, here's a brief overview. *The Richest Man in Babylon* is a classic personal finance book written by the American industrialist, businessman, and (eventually) author George S. Clason. The book was first published in 1926, although the content was initially distributed through pamphlets and only later compiled into a book. It has been reprinted countless times in the last 100 years, and many swear it is the greatest personal finance book ever written.[18]

As its title hints, this book compiles a series of parables set in ancient Babylon, focusing on financial wisdom and strategies for achieving personal wealth and financial security. The book speaks to the importance of financial independence and preaches wealth accumulation through disciplined saving, prudent spending, and wise investing. The lynch pin of the book is a series of (fictitious) Babylonian tablets located by a so-called Professor Caldwell, which were sent for deciphering to the Department

[17] Relevant, again, here also is the work by Bellhouse (2017), and see also Ciecka (2012) for more on Ulpian and his tables.

[18] See, for example, the fawning article, which is a review of the book and restatement of its relevance, nearly a century after it was published, in *Forbes Magazine*. https://www.forbes.com/sites/rickferri/2010/09/01/the-richest-man-in-babylon/?sh=2dec4fa6713d.

of Archaeology in Nottingham, UK.[19] Of course, the ration tablet I noted in this chapter are from real (not fictional) archeological discoveries. Interestingly, the book does have something to say about retirement planning, under the title—the sixth of a total of seven cures for a lean purse—labelled *insure a future income*. George S. Clason writes:

> No man can afford not to insure a treasure for his old age and the protection of his family, no matter how prosperous his business and his investments may be…Provide in advance for the needs of thy growing age and the protection of thy family.[20]

And, while there is no mention of insurance annuities or longevity risk pooling—since the action takes place in ancient Babylon—there is one statement that is rather apropos to this book. The protagonist, the "rich man" from Babylon named Arkad, notes the following lesson to his class:

> In my mind rests a belief that some day wise thinking men will devise a plan to insure against death whereby many men pay in but a trifling sum regularly, the aggregate making a handsome sum for the family of each member of passeth [*sic*] to the beyond. This do I see as something desirable and which I could highly recommend. But, today it is not possible because it may reach beyond the life of any man or any partnership to operate. It must be as stable as the king's throne. Some day I feel that such a plan shall come to pass and be a great blessing to many men, because even the first small payment will make available a snug fortune for the family of a member should he pass on.[21]

Conclusion

In the beginning, in a time before stocks or even bonds, there were pension life annuities. It might come as a surprise that life annuities could be found in Babylonian times, and predate the trading of bonds let alone stocks. Long before British, French, Dutch, or German governments

[19] Clason was born in Louisiana, moved to Colorado and died in California, all of them in the US. Presumably the British address for the Babylonian tables added gravitas and authority.

[20] Clason (1926, p. 38).

[21] Clason (1926, p. 39).

borrowed through fixed-maturity bonds, they financed deficits using life contingent instruments. Indeed, the history of life annuities spans thousands of years, predating common stocks, savings bonds, mutual funds, and the exchange of traded funds. While today only insurance companies can issue life annuities, in the past, churches, cities, and states were known to issue both life annuities and tontines.[22]

As noted in this chapter, an account in the Hebrew Bible tells the story of a life annuity granted to Jehoiachim, King of Judah, by the Babylonian king in 562 BC following his release from prison. And, I used the Roman jurist Ulpian's table to value that pension annuity, since life annuities gained traction in Rome during 100 AD and 200 AD, with mutual aid societies of the Roman Legions providing them to retiring soldiers.

For a millennium, monasteries and churches were the primary traders of life annuities. A notable transaction occurred in 1308 when the Abbot of St. Denis, near Paris, sold an annuity to the Archbishop of Bremen for 2400 livres, promising an annual return of 400 livres. This deal, yielding 16.66%, was later contested by the abbot as usurious, despite its initial acceptance. Although I didn't focus on this aspect of life annuities, they were often seen as a way to earn interest without breaching usury laws, as the mortality risk involved lent the transactions a gambling-like aspect, which was not forbidden.[23] By the thirteenth century, cities in regions like Flanders and Brabant (modern-day Belgium) began issuing life annuities. By the sixteenth century, this practice had expanded to cities and governments using annuities to fund deficits. In 1554, for instance, the Dutch Republic financed a deficit by selling life annuities worth 100,000 guilders. Annuities had became ubiquitous.[24]

Yet, to conclude once again with the main point, annuities might have been available in Babylonian times, but the first-ever attempt to pool and eliminate the idiosyncratic risk took place a mere 280 years ago, in Edinburgh, in the year 1744.

[22] Much more (history) in Milevsky (2013,2015).

[23] Ibid.

[24] See the book by Prak and van Zanden (2023, p. 84).

Sources and Additional References

Alstola, T. (2019). Chapter 2 Judean Royalty and Professionals in Babylon. In *Judeans in Babylonia: A Study of Deportees in the Sixth and Fifth Centuries BCE*. Brill.

Beaulieu, P. A. (2018). *A History of Babylon, 2200 BC-AD 75*. Wiley.

Begg, C. T. (1986). The Significance of Jehoiachin's Release: A New Proposal. *Journal for the Study of the Old Testament, 11*(36), 49–56.

Bellhouse, D. R. (2017). *Leases for Lives: Life Contingent Contracts and the Emergence of Actuarial Science in Eighteenth-Century England*. Cambridge University Press.

Clements, R. E. (2007). A Royal Privilege: Dining in the Presence of the Great King (2 Kings 25.27–30). In *Reflection and Refraction*. Brill.

Clason, G. S. ([1926] 1955). *The Richest Man in Babylon*. Signet Book.

Ciecka, J. E. (2012). Ulpians Table and the Value of Life Annuities and Usufructs. *The Journal of Law and Economics, 19*, 7.

de Vries, T., & Zwalve, W. J. (2004). Roman Actuarial Science and Ulpian's Life Expectancy Table. In *Roman Rule and Civic Life: Local and Regional Perspectives* (pp. 275–297). Brill.

Feldman, L. H. (1995). Josephus' Portrait of Jehoiachin. *Proceedings of the American Philosophical Society, 139*(1), 11–31.

Finegan, J. (2015). *The Handbook of Biblical Chronology*. Hendrickson Publishers.

Gillman, B. M. (2014). Review of Life Annuities: An Optimal Product for Retirement Income. *Financial Analysts Journal, 70*(1), 68–70.

Graeber, D. (2011). *Debt: The First Five Thousand Years*. Melville House Publishing.

Grellet, R. (2018). *A Tablet, a King and His Rations*. Armstrong: Institute of Biblical Archeology.

Homer, S., & Sylla, R. E. (2005). *A History of Interest Rates* (4th ed.). Wiley.

Horn, S. H. (1967). The Babylonian Chronicle and the Ancient Calendar of the Kingdom of Judah. *Andrews University Seminary Studies (AUSS), 5*(1).

Knauf, E. A., & Guillaume, P. (2016). *A History of Biblical Israel: The Fate of the Tribes and Kingdoms from Merenptah to Bar Kochba* (p. 137). Equinox.

Kopf, E. W. (1927). *The Early History of the Annuity*. LW Lawrence.

Lewin, C. G. (2003). *Pensions and Insurance Before 1800: A Social History*. Tuckwell Press.

Milevsky, M. A. (2013). *Life Annuities: An Optimal Product for Retirement Income*. CFA Institute Research Foundation Monograph.

Milevsky, M. A. (2015). *King William's Tontine: Why the Retirement Annuity of the Future Should Resemble Its Past*. Cambridge University Press.

Neufeld, E. (1954). The Rate of Interest and the Text of Nehemiah 5.11. *The Jewish Quarterly Review, 44*(3), 194–204.

Prak, M., & van Zanden, J. L. (2023). *Pioneers of Capitalism: The Netherlands 1000–1800*. In J. Mokyr (Ed.), *Part of the Princeton Economic History of the Western World*. Princeton.

Sack, R. H. (2004). *Images of Nebuchadnezzar: The Emergence of a legend* (2nd Revised and Expanded Ed.). Susquehanna University Press.

Thiele, E. R. (1956). New Evidence on the Chronology of the Last Kings of Judah. *Bulletin of the American Schools of Oriental Research, 143*(1), 22–27.

Webster, A. (1740). *The Wicked Life, and Fatal But Deserved Death of Haman, Ahasuerus's Prime Minister*, a Sermon Preached before the Right Honourable Lord Provost, Magistrates and Town-Council of Edinburgh, at their Election, on Tuesday, 30th September, 1740.

Webster, A. (1746). *Heathens Professing Judaism, When the Fear of the Jews Fell Upon Them.* The Substance of Two Sermons Preached in the Tolbooth Church, Edinburgh, On Occasion of The Thanksgiving, June 23rd, 1746 appointed by the late General Assembly of the Church of Scotland, for the Victory obtained over the Rebels at the Battle of Culloden, April 16th, 1746.

Weiershäuser, F., & Novotny, J. (2020). *The Royal Inscriptions of Amēl-Marduk (561–560 BC), Neriglissar (559–556 BC), and Nabonidus (555–539 BC), Kings of Babylon* (Vol. 2). Penn State Press.

PART III

From Retirement to Decumulation

CHAPTER 11

Pension Resistance in the Nineteenth Century

Introduction & Objectives

As first noted in Chapter 1, and alluded to a few times over the course of this book, the German chancellor Otto van Bismarck introduced the first national pension scheme, which was enshrined by the Reichstag in 1889. That was a very important milestone in pensions history and this is what the professional historians have to say about the matter. In an article published in the *American Historical Review* in 2018, by Erik Grimmer-Solem, which itself was a review of the 2016 book entitled *A History of the German Pension System* by A.C. Mierzejewski, both the reviewer and author crowned the German state pension as the first system of its kind in the entire world.[1] It was only after Germany implemented a widespread state pension that the United Kingdom of England, Scotland, and Ireland introduced its own national pension scheme in the early twentieth century under the leadership of Herbert H. Asquith. That was formally known and referred to today as the Old-Age Pensions Act of 1908.[2] Then, a few decades later in the United States, Social Security was enacted as part of Franklin D. Roosevelt's New Deal in 1935. The history of Social Security is described at length on the website of the US Social Security administration, but here I simply note that it too credits the beginning to the

[1] Source: E. Grimmer-Solem (2018, pp. 318–320).

[2] *The Old-Age Pensions* Act (1908).

M. A. Milevsky, *The Religious Roots of Longevity Risk Sharing*,
https://doi.org/10.1007/978-3-031-62403-2_11

year 1889 and to the (Protestant) Chancellor Bismarck.[3] Now, Bismarck's reasons and motivations have been the subject of much historical debate, and researchers have noted that he was rather hostile to capitalism, and private industry was equally hostile to some of his ideas, as I'll discuss in a moment.[4] The only thing I would remind readers about at this point is that the actuarial methodology and institutional infrastructure for offering pension life annuities had its genesis over 150 years prior, in Edinburgh, Scotland.

Indeed, over a century after Bismarck's innovation, almost all developed countries around the world now have some form of national pension scheme that effectively provides a retirement annuity to older citizens, often as part of a broader welfare state. Of course, that is why today, in the year 2024, Mercer consulting and the CFA Institute (see Chapter 1) are able to rank national pension schemes around the world. In fact, Article 25 of the 1948 Declaration of Human Rights states, "Everyone has the right to a standard of living…and the right to security in the event of…old age."[5] So, as I explained earlier in Chapter 1 and in the introduction of this book, the concept of a guaranteed state pension has gone from completely non-existent three centuries ago to a financial fact that is ubiquitous. Old people around the world are entitled to some sort of retirement pension or life annuity, and the only question is how generous, how sustainable, and how widespread these pensions and annuities are within society. To emphasize their ubiquity, Table 11.1 (which includes many countries not ranked by the Mercer CFA Institute Global Pension Index) is a hand-collected list of over 50 countries with national pension schemes, showing the date when the legislation was instituted and the name of the local champion or advocate. So, although there continues to be a robust global debate in the financial, economic, and actuarial literature on exactly how to design a sustainable national pension scheme, or how to improve a country's Mercer CFA Institute Pension Index score, even the most ardent *individualists* and free market advocates do not question the need for some government

[3] see link https://www.ssa.gov/history/briefhistory3.html#:~:text=The%20Social%20Security%20Act%20was,a%20continuing%20income%20after%20retirement.

[4] See Chapter 7, authored by Colin Stewart, in the collection of article edited by Derek Renn, cited as Renn (1998).

[5] See the 1948 Declaration of Human Rights. https://www.un.org/en/udhrbook/pdf/udhr_booklet_en_web.pdf.

provision of retirement annuities.[6] For example, Social Security is widely considered by both Republicans and Democrats in the United States as one of the most popular government programs, which is rather surprising considering these two disparate political groups rarely agree on anything.[7]

What seems to have received less attention than the broad-based and global nature of pensions is the disparate length of the history as opposed to the quality. Assuming that "the world" learned how to pool, manage, and price longevity risk in the mid-eighteenth century, as argued in Part 2 of this book, this chapter asks: *why and when*: Why did it take over 100 years for national pension schemes to take root around the globe? Why did some countries and nation-states adopt them much faster than others? Why did Germany implement this 50 years before the United States? And, while a few decades might not seem like a long time to wait from the perspective of the twenty-first century, why did some nation-states provide retirement annuities while others delayed them? Even today some countries continue to struggle with fragmented and regional schemes.[8] Although the Protestant Christian religious origins can explain some of this, there must be more to the story.

Now, perhaps the kneejerk reaction—among trained economists—is to embed or graft pension provision into one of the chapters of long-run economic development and to argue that higher levels of gross domestic product (GDP) lead to greater pension provisions.[9] So, for example, European countries like Germany (in 1889) were more likely to implement national pensions before Ethiopia, which finally did so in 1963. After all, one can't expect a country or region that is in the midst of a civil war or struggling with the provision of clean water and sanitation to worry about pension and life annuities. Yet, while not dismissing the slow, uneven, and rocky pace of development as leading to the same unevenness for pensions, there are many exceptions to this "don't put the pension cart before the horse" rule of thumb. In fact, one of the starkest examples showing that economic development is there but national pensions are

[6] See the 1994 report by the World Bank in Washington D.C., published by Oxford University Press, and titled: Averting the Old Age Crisis: Policies to Protect the Old and Promote Growth.

[7] See article "Social Security and American Public Opinion" by Frank Newport, published June 18, 2019 by Gallup.

[8] See C. Arza and P. Johnson (2006).

[9] See for example T. Picketty (2014).

Table 11.1 Global pensions heroes and beginnings

Country	*Year*	*Champion/President*
Germany	1881	Bismarck
Denmark	1891	King Christian IX
New Zealand	1898	Richard Seddon
Argentina	1904	Julio Roca
Austria	1906	Theodor Korner
Australia	1908	Andrew Fisher
The UK	1908	David Lloyd George
Ireland	1908	David Lloyd George
France	1910	Armand Fallieres
Sweden	1913	Anders Lindstedt
Italy	1919	Vittorio Orlando
Netherlands	1919	Piet Aalberse Sr
Spain	1919	Pablo Iglesias
Cuba	1920s	Alfredo Zayas
Russia	1922	Vladimir Lenin
Brazil	1923	Arthur Bernardes
Chile	1924	Arturo Alessandri Palma
Poland	1927	Ignacy Moscicki
South Africa	1928	James Barry Munnik Hertzog
Canada	1927	Mackenzie King
Greece	1934	Panagiotis Kanellopoulos
US	1935	Franklin Roosevelt
Norway	1936	Johan Nygaardsvold
Finland	1939	Kyosti Kallio
Japan	1941	Tojo Hideki
Costa Rica	1941	Rafael Calderon Guardia
Mexico	1943	Manuel Avila Camacho
Turkey	1949	Ismet Inonu
China	1951	Mao Zedong
India	1952	Jawaharlal Nehru
Singapore	1953	David Marshall
Philippines	1954	Ramon Magsaysay
Egypt	1955	General Naguib
Israel	1957	David Ben-Gurion
South Korea	1960	Yun Posun
Tunisia	1960	Habib Bourguiba
Nigeria	1961	Nnamdi Azikiwe
Saudi Arabia	1962	Abdulaziz Saud
Gabon	1963	Léon M'ba (president)
Ethiopia	1963	Aklilu Habte-Wold
Switzerland	1963	Willy Spühler
Ghana	1965	Kwame Nkrumah

(continued)

Table 11.1 (continued)

Country	*Year*	*Champion/President*
Kenya	1965	Jomo Kenyatta
Jamaica	1966	Alexander Bustamante
Pakistan	1972	Zulfikar Ali Bhutto
Hong Kong	1973	Sir Murray MacLehose
Thailand	1990	Chatichai Choonhavan
Hungary	1998	Viktor Orban

lacking is in the United States, which lagged behind Germany by almost a half-century.

Political scientists and economically trained demographers might answer the above-noted *why and when* question by appealing to non-economic considerations such as ageing societies, institutional stability, and especially political forces, all of which have much merit. In Germany, for example, Bismarck originally sought a much more centralized and generous scheme. However, after 43 Reichstag committee meetings and 24 plenary sessions, even the "Iron Chancellor" had to settle for a watered-down version because of the opposition from both left-wing liberals and social democrats.[10]

The approach I would like to take in answering the *why and when* research question is one that paraphrases Newton's third law of motion, namely, every attempt to introduce a national pension scheme has faced an opposing force.

Forces: For & Against

Returning to the central question of *why and when*, I will now tackle the matter somewhat differently by focusing on neglected groups within the history of national pension schemes, made up of *opposing business interests*. In particular, I would like to discuss the global economic force that stood in their way both overtly and covertly. These groups were a loose confederation of commercial insurance companies, local savings banks, and provident societies, collectively named the "commercial resistance." They were well aware of how insurance risk could be pooled, in fact

[10] Source: A.C. Mierzejewski (2016), E. Grimmer-Solem (2018) as well as U. Haerendel (2009).

many of them implemented such schemes on smaller local levels themselves. But they combated the larger national proposals, lobbied against the creation of larger risk pools, and impeded the political progress of pension planning programs.

Why stand in the way? Because these programs would reduce consumer demand for the saving, investment, and life annuity services these groups offered, which was their business *raison d'être*. Eventually, of course, these opposing business forces lost and they learned to coexist with national competition for their private annuities. These opposing forces varied in strength around the world. Hence, in some national regions, they were more successful in resisting pension schemes compared to others. I will try to convince you of this thesis over the next few pages.

Moreover, even after defeat, these opposing commercial forces continued to exert a powerful influence during the growth and evolution of national pension schemes. In this chapter, I will review the rationale for objections, discuss the historiography, and provide some examples from Germany, the UK, and the United States. The discussion also provides a basis for a secondary thesis that national pension schemes might have emerged a century earlier—closer to the eighteenth-century work described in Part 2 of the book —had it not been for the push back by global commercial forces.

Pension Scheme Classification

National pension schemes are part of a larger category of government-run social insurance programs and can be viewed as the third leg of an economic stool consisting of health insurance, accident insurance, and, finally, retirement life annuities. One can trace the origin of almost all social insurance programs (at least in the UK) to England's Poor Relief Act of 1598, also known as the Elizabethan poor laws.[11] In this chapter, I will narrowly and exclusively focus on the state provision of retirement annuities, the third leg of the stool, which is the most obvious component of social insurance programs that might compete with the private sector. More on this to follow.

[11] See L. A. Botelho (2004), as well as well as R. Houlbrooke and L. A. Botelho (2004), and Boyer, G. R. (1990), and more generally the treatment of old age in P. Thane (2020).

Within the context of national pension schemes that offer these retirement annuities to older citizens, as I had alluded to in the introduction, there are two methods or funding schemes by which to finance and pay for these social programs. The first is via the national tax system in which those who are currently working (i.e., contributing) implicitly pay the pensions of those who are currently retired (i.e., drawing down). This is often labelled as an *unfunded* pay-as-you-go (PAYGO) scheme, because there is no separate or distinct pool of money that is held exclusively to pay out these retirement annuities. These unfunded PAYGO schemes—which were quite popular in the mid-twentieth century—now tend to be criticized by pension experts and penalized by "league rankings," such as the one described in Chapter 1. The PAYGO scheme can work (in theory) when (young) populations are growing and/or when the number of older retirees is shrinking, but not the other way around.

In contrast to PAYGO schemes, the second method is one in which participants contribute to a communal fund while they are working, which then serves as the pool from which money is withdrawn during their retirement. That is (much) closer to the Presbyterian annuity scheme discussed in the earlier part of this book, since ministers and professors contributed to a fund that grew over time and was actuarially calculated to "have enough" to pay the widows and orphans.

This second financing method creates a tighter link between savings (while working) and benefits (while retired) and is often called a contributory system, for obvious reasons. Not to get bogged down in minutia, but a contributory system can be funded by employers as well as employees and perhaps even by government. It can be tightly linked to wages in which retirees are guaranteed, for example, 2% of their salary per year of service. So, after 30 years of work, they might receive a retirement annuity of 60% of their final salary. Hopefully, it's clear to readers that these national pensions are very similar to life annuities that are available in the private sector and that can be acquired or purchased via a lump sum or in instalments.

All of this is by no means a proper or full description of the spectrum of national pension schemes, but is the minimum needed for what follows.[12] The key takeaway here is the similarity between retirement annuities and those available in the private sector. Theoretically, one can save money in

[12] For a full and proper description see the book by David Blake, in his (2006) book titled: *Pension Economics*.

a simple bank or investment account and purchase the same annuity from the private sector. This is (and was) precisely the source of commercial objection, unlike paved roads, hospitals, and the military, which are quite difficult, if not impossible, to contract privately. The national pension scheme was in direct competition.

Pension Origins & Objections

Insurance actuaries and pension economists who study the financing of old age have long noted the ancient pedigree of retirement annuities that provide a lifetime of income during old age. As I had mentioned in Chapter 10, two thousand years ago, veterans of the Roman army were granted pension annuities for honourable service. Once again, the very early attempt at pension funding calculations, sometime around AD 200, by the Roman jurist Ulpian was based on the age of the retiree.[13] Here, I just remind readers about the classical references to state and government retirement annuities even earlier than Roman times. For example, in 400 BC, the Greek orator Lysias complained when his retirement annuity was discontinued in his old age, and Chapter 10 already told the story of Jehoiachin, King of Judah, who was granted a lifetime pension from Amel-Marduk, King of Babylon.[14]

As noted earlier, pensions are credited to Chancellor Bismarck, but in fact the idea of a national plan was first proposed over a century before by Marquis de Condorcet (b. 1743, d. 1794), who was French. In the posthumously published *Esquisse d'un tableau historique des progres de l'espirit humain,* he advocated creating a national pension scheme that would grant retirement annuities to the elderly that took advantage of recent advances in statistics and probability[15]:

> There exists then a necessary cause of inequality, of dependence, and even of penury, which menaces without ceasing the most numerous and active

[13] See the chapter entitled "Ancient Civilizations," in the book by Chris G. Lewin called: *Pensions and Insurance Before 1800,* which was published in 2003.

[14] Source: *The Bible. Authorized King James Version with Apocrypha.* 2 Kings. Chapter 25. Verse 27. See also Lewin (2003).

[15] Source: M. J. Condorcet (1955). See also W. A. Darity, (2008) Jr. (Ed.), *International Encyclopedia of the Social Sciences* (2nd ed., Vol. 2, pp. 65–66). Macmillan Reference.

> class of our societies. This inequality, however, may be in great measure destroyed, by setting chance against chance, in securing to him who attains old age, a support, arising from his savings, but augmented by those of other persons who making a similar addition to a common stock, may happen to die before they shall have occasion to recur to it.[16]

Resistance appeared soon after Condorcet's proposal in the form of eighteenth-century English economist and historian Thomas Robert Malthus (b. 1766, d. 1834). He wrote, consistent with his worldview, that:

> Were every man sure of a comfortable provision for a family, almost every man would have one. And, where the rising generation free from the killing frost of misery, population must rapidly increase.[17]

The incentive effects raised by Malthus were soon echoed by other economists who went well beyond the cliché and bleak concerns about overpopulation and scarce resources, which is commonly associated with Malthusian views. The intellectual focus then shifted to insights of the Italian Nobel laureate in economics, Franco Modigliani (b. 1900, d. 1990), and his so-called lifecycle hypothesis of saving and investing. In the 1950s, Modigliani theorized that consumers with a limited working life, but a longer biological life span, would save and spend in such a way to smooth their consumption and standard of living over their entire life. In their working years they would rationally set aside ample provisions for old age and would spend those provisions once they were no longer working. This framework is known as the lifecycle model[18] and has been the foundation of "rational microeconomic theory" for over half a century. It is echoed in Milton Friedman's permanent income hypothesis,[19] which is predicated on the idea that consumers rationally spread resources over their working life. Citizens can create their own pensions.

[16] Source: M. J. Condorcet (1955).

[17] Sketch for Historical Picture of the Progress of the Human Mind, trans. *June Barraclough. London: Weidenfeld & Nicolson.*

[18] Condorcet, Marquis de., entry in Darity (2008).

[19] This is included in (what is likely) one of the most famous books in modern economics, written by Milton Friedman in 1957, called *A theory of the consumption function*.

The relevant point here, as it is related to national pension scheme opposition, is as follows: If governments were going to step in and offer retirement annuities to individuals above a certain age, individuals who would have no doubt planned for their old age by saving during their working years would compensate by saving less themselves. Instead of depositing funds in a local bank, participating in a community-based benevolent society fund, or investing in a private life annuity sold by an insurance company, they would instead participate in the new government scheme. Or, stated differently, the national pension scheme would crowd out the private sector. And, while Modigliani and Friedman and their theories were not around or alive in the late nineteenth and early twentieth century, their ideas were intuited and echoed by commercial interests in the private sector. While this "theory" that longevity risk pooling would have emerged earlier were in not for the objections from industry is obviously a statement about counterfactuals, I will provide a few case studies and some economic literature that is consistent with that view. There was a crowding out effect, as the commercial groups suspected.

National Examples: Germany, the UK, and the US

As Germany was the first country to implement a national pension scheme, it is the first place to look for commercial resistance. Two historians writing in the *European Review of Economic History* allude to the Modigliani-Friedman argument and document that this concern was more than a theoretical fear. It actually happened. Many consumers reduced their retirement savings in response to the new government program.

Using data on private savings deposited in savings banks within 437 Prussian counties during the last two decades on the nineteenth century, the historians measured the extent of this effect. In their words, "We clearly see the expected downward trend in the group's relative savings activities after the introduction of the Bismarck's social insurance system." Furthermore, "our regression analysis suggests that the introduction of social security for large parts of the Prussian population [in the 1880s] crowded out private savings."[20] At the risk of a teleology, their research

[20] Lehmann-Hasemeyer and Streb (2018). See also Scheubel (2013).

results are consistent with precisely this crowding out effect that was feared by commercial interests a full century earlier.

Thus, while debates in the late nineteenth century weren't couched in the language of the above-noted lifecycle model of the mid-twentieth century, it is quite clear that those who objected intuited and grasped the ideas of Modigliani and Friedman, a generation before they were born. Namely, if consumers are forced to save money for their old age, they will reduce their discretional savings—thus harming the business and commercial interest of those who manage those savings.

Moving from Germany to the UK, we find that concerns were similar. In an article published in the 1960s in the *Economic History Review*, the author, B. B. Gilbert, states: "In the two decades during which old age pensions had received serious political consideration, the proposal was strenuously resisted by...the industrial insurance companies and friendly societies, which saw the government as a competitor in the field of public welfare."[21] According to Gilbert's article, the most strenuous opposition to old-age pension legislation came from *friendly societies* who represented the British working man (in his words) and their provision for domestic emergencies. Friendly societies epitomized the Victorian ideals of thrift, respectability, individual responsibility, and self-help. These groups offered a sickness benefit and financial assistance that provided income to members who were old, infirm, disabled, and therefore unable to work; that is, they offered a pension of sorts. To Gilbert's point, in the year 1896, one of those friendly societies—the Independent Order of Rechabites (IOR)—with over a quarter of a million members, surveyed the membership and reported that 99% of them "were against any form of compulsory pension contribution." A letter sent to the 1896 Rothschild Committee stated: "If we cannot manage this better than the state, we had better shut up shop altogether."[22]

What was the Independent Order of Rechabites? It was a friendly society founded in England in 1835 as part of the wider temperance movement that was sweeping through the country at the time. The IOR was established by members of the temperance movement who were concerned about the social and individual problems caused by alcohol consumption. The Rechabites—wait for it here—were named after the

[21] B.B. Gilbert (1965).

[22] Source: B.B. Gilbert (1965, p. 560).

biblical tribe of Rechabites whose members abstained from wine and aimed to promote temperance and moral values through personal example and mutual support. As noted above, the IOR provided its members with insurance and mutual aid benefits, which was a common practice among such societies in the nineteenth and early twentieth century. These benefits included financial support in times of sickness, unemployment, or death, and most importantly offered pensions and life annuities.

In case you were wondering, the IOR was indeed rooted in Protestant Christian principles, reflecting the temperance movement's strong ties to Protestantism. Many temperance organizations, including the Rechabites, were founded on Christian ethics and morality, emphasizing virtues such as sobriety, self-control, and piety. They often held meetings that resembled religious services, including prayers, hymns, and sermons promoting abstinence from alcohol. So, to be very clear, while the IOR was not a church or a solely religious organization, its foundation and activities were heavily influenced by Protestant Christian ideals. It operated within the framework of promoting moral and ethical living according to Protestant values, using the narrative and example of the biblical Rechabites to inspire and justify its temperance mission.

Back to national pensions, the initial report examining the feasibility of these plans in the UK concluded: "The friendly societies would not permit contributory pensions" and despite the fact that everyone approved of the idea of a pension in principle, "there would be no pensions." A National Committee summarized the situation as follows: "It is impossible to conceive of any plan by which the contributions can be drawn from the masses of people alongside the friendly society contributions."[23]

I hope you can see the irony of some Protestant Christian organizations objecting to national longevity risk pooling schemes. One can only speculate if Reverend Alexander Webster, who clearly was not and never would be a member of the temperance movement, would have objected to a national (Scottish) system that would possibly compete with the successful church scheme.

What changed? Why did the opposition finally acquiesce? Well, it seems the managers of the largest 34 friendly societies, representing over 2.3 million members in 22,000 separate branches, hadn't quite mastered the Marquis de Condorcet's mathematical lessons of setting "chance against

[23] Source B.B. Gilbert (1965, p. 562).

chance." According to Gilbert's article, by the early twentieth century, the friendly societies who fiercely opposed pensions were now faced with a problem that "they could not solve and which they scarcely understood."[24] Namely, a large number of their members were no longer able to work due to old age, but at the same time "did not die." They were facing bankruptcy and insolvency. By 1908 they had no choice but to accept a national old-age pension system. In other words, they had certainly not learned the actuarial mathematics and lessons of Alexander Webster's annuity scheme.

Moving across the Atlantic, the dialogue and rhetoric in the United States prior to the creation of social security in 1935 was quite similar. Writing in 1910 in the *Quarterly Journal of Economics*, which was and continues to be the premier research outlet for scholarly economists, F. S. Baldwin expressed similar concerns about the incentive effects of national pensions: "If the state granted gratuitous pensions for old age, the fact would doubtless be taken into account by workers."[25] In other words, they would adapt their own saving and spending behaviour. He notes, "The amount of the pension would be discounted in advance by the workers in their competition for employment." And, with regard to non-contributory unfunded pension schemes, Baldwin writes: "The motives and energies of self-help would be weakened by this form of state help."

The negative opinion about national pensions continues with: "It threatens disaster to voluntary agencies for the encouragement of saving, such as friendly societies." Baldwin is so convinced that he writes, "It seems so obvious that it is surprising to find among professional economists any dissent on this question." Lest this be a purely economic argument, the forces of pension resistance were more than an arm of commercial, business, or capitalistic interests. In this opinion, "pension policy would weaken the bonds of family solidarity. It would take away in part the filial obligation for support of aged parents which is one of the main ties that hold the family together."[26]

In yet another perverse incentive created by national pension schemes in the United States, it seems a number of employers took advantage of newfound government largesse. Writing about the history of pensions in

[24] Gilbert (1965, p. 566).

[25] Baldwin, F. S. (1910).

[26] See Baldwin (1910, specifically p. 726).

America, economic historians R. Clark and L. A. Craig have concluded, "The creation of the Social Security system was seen by employers as a subsidy for their retirement plans. Instead of being responsible for bearing the full weight of the worker's old-age income, employers could point to future social security as a component of their employees' income replacement rate."[27]

There is yet another aspect to this matter. Even when countries implement old-age pension schemes, the technical parameters of the pension system depend on another global force: political regimes. In an intriguing article published in *Comparative Political Studies*, A. Grünewald makes the rather provocative claim that while old-age pension systems are now ubiquitous in over 120 different countries around the world, there is a marked difference based on political regimes.[28] It seems that democratic regimes have higher pension coverage rates; that is, the fraction of the population that is part of the pension system. Whereas non-democratic regimes have higher pension replacement rates, offering more to retirees.

Conclusion

National pension schemes are widely accepted as part of basic human rights but were introduced and adopted at varying times in recent history, as Table 11.1 shows. The difference in adoption times, spread over a century, was not just a matter of demographics or politics, but had much to do with opposing global forces that sought to maintain their dominance and control in the industry.

To conclude, Robin Blackburn, in his 2002 book entitled *Banking on Death or Investing in Life: The History and Future of Pensions*, notes with much concern—and perhaps a touch of irony—that after a full century, national (as well as corporate) pension schemes have yielded much ground to the interests of the private sector. The same participants, that is, the insurance companies, financial cooperatives, and mutual organizations that objected to the crowding out created by governments, have all lined up to reap the benefits of privatization. In fact, the monies collected by government as part of the pension funding process, often through

[27] Clark, R. and Craig, L. A. (2018).

[28] Grünewald, A. (2021).

the tax system, is being managed and administered by commercial interests in the private sector. This is the same global force that fought (and lost) the battle during the last century. Clearly, there are many other forces at work—beside religion—in the historical evolution of pensions and longevity risk pooling. But there is yet another social force at play, which is the topic of the next chapter.

Sources and Additional References

Arza, C., & Johnson, P. (2006). The development of public pensions from 1889 to the 1990s. In G. L. Clark, A. H. Munnell & J. M. Orszag (Eds.), *The Oxford handbook of pensions and retirement income*. Oxford University Press.

Baldwin, F. S. (1910). Old age pension schemes: A criticism and a program. *The Quarterly Journal of Economics, 24*(4), 713–742.

Blake, D. (2006). *Pension economics*. John Wiley & Sons.

Botelho, L. A. (2004). *Old age and the English poor law of 1500–1700*. Boydell Press.

Boyer, G. R. (1990). *An economic history of the English poor law: 1750–1850*. Cambridge University Press.

Clark, R., & Craig, L. A. (2018). Retirement and pensions in American economic history. In L. P. Cain, P. V. Fishback, & P. W. Rhode (Eds.), *The Oxford handbook of American economic history*. Oxford University.

Condorcet, M. J. (1955). *Sketch for a historical picture of the progress of the human mind* (trans. June Barraclough). Weidenfeld & Nicolson.

de Condorcet, M. (2008). In W. A. Darity, Jr. (Ed.), *International encyclopedia of the social sciences* (2nd ed., Vol. 2, pp. 65–66). Macmillan Reference.

Friedman, M. (1957). The permanent income hypothesis. In *A theory of the consumption function* (pp. 20–37). Princeton University Press.

Gilbert, B. B. (1965). The decay of nineteenth-century provident institutions and the coming of old age pensions in Great Britain. *The Economic History Review, 17*(3), 551–563.

Grimmer-Solem, E. (2018). Review of A. C. Mierzejewski. A history of the German public pension system: Continuity amid change. *The American Historical Review, 123*(1), 318–320.

Grünewald, A. (2021). From benefits and beneficiaries: The historical origins of old-age pensions from a political regime perspective. *Comparative Political Studies, 54*(8), 1424–1458.

Haerendel, U. (2009). Ideas of justice in the foundation of the German old age pension system. *GHI London Bulletin*, Bd. 31, Nr. 2.

Houlbrooke, R., & Botelho, L. A. (2004). Old age and the English poor law, 1500–1700. *The American Historical Review*, 111(2), 560–561.

Lehmann-Hasemeyer, S., & Streb, J. (2018). Does social security crowd out private savings? The case of Bismarck's system of social insurance. *European Review of Economic History, 22*(3), 298–321.

Lewin, C. G. (2003). *Pensions and insurance before 1800.* Tuckwell.

Malthus, T. R. (1872). *An essay on the principle of population* (Chapter 8, p. 65). Oxford World's Classics.

Mierzejewski, A. C. (2016). *A History of the German public pension system: Continuity amid change*. Lexington Books.

Modigliani, F., & Brumberg, R. (1954). Utility analysis and the consumption function: An interpretation of cross-section data. *Franco Modigliani, 1*(1), 388–436.

Picketty, T. (2014). *Capital in the twenty-first century* (pp. 488–490). The Belknap Press of Harvard University Press.

Renn, D. (Ed.). (1998). *Life, death and money: Actuaries and the development of social and financial markets*. John Wiley & Sons.

Scheubel, B. (2013). Bismarck's pension system. In *Bismarck's institutions: A historical perspective on the social security hypothesis* (pp. 77–104). Mohr Siebeck GmbH.

Thane, P. (2020). Old age in European cultures: A significant presence from antiquity to the present. *The American Historical Review, 125*(2), 385–395.

The Old-Age Pensions Act. (1908, October 15). *The text of the regulations made thereunder* (3rd ed.). Official Circulars and Instructions for the Guidance of Pension Authorities by the Local Government Boards of England, Scotland and Ireland. Chas. Knight & Co., Ltd. Internet Archive

CHAPTER 12

Longevity Heterogeneity in the Twenty-First Century

One of the arguments that I have made in Part II of this book, and specifically in Chapters 6, 7, and 8, was that the success of the first longevity pooling scheme in the mid-eighteenth century was partially driven by elements of (ethical) solidarity and (religious) affinity among the ministers. In particular, recall the reply that Reverend Alexander Webster sent, in response to complaints—and newspaper editorials—that the elderly ministers who were more likely to die sooner were selecting the higher annuity benefits. His response, in a nutshell, was that the younger brethren were aware and fully acquiesced to this "anti-selection." Why did they agree to pay (contribute) more than they should and receive (a multiple) less than they should? For the sake of the greater good and a brotherly spirit. Recall also some of the objections to larger national pension plans, in Chapter 11, expressing concern they might disincentivize effort and work. So, in this final chapter I will tackle an issue that has been brewing under the surface of longevity risk pooling ever since I introduced Colin and Anne in Chapter 3. What happens in a longevity risk pool when it becomes quite clear that some "swimmers" are benefiting much more than others?[1]

[1] Part of this (and the final book's) chapter echoes and continues themes discussed in an article that I wrote entitled: "Swimming with wealthy sharks: longevity, volatility and the value of risk pooling," which was published in the *Journal of Pension Economics*

M. A. Milevsky, *The Religious Roots of Longevity Risk Sharing*,
https://doi.org/10.1007/978-3-031-62403-2_12

There is a growing and alarming body of evidence telling us that longevity isn't as random as the coin toss—and the discussion within Chapter 2 alluded to these disparities. Whether it be genetics, environment, lifestyle choices, or religious beliefs, longevity prospects (or curves) simply aren't the same for everyone in society. For the less fortunate groups, there is little chance they will live to an advanced age and benefit from the insurance aspects of longevity risk pooling. Within society, these groups have a legitimate claim that national pension schemes that pool everyone together and pay out the same to everyone, aren't fair or equitable.

In fact, in early 2023, a group representing Aboriginal Australians filed a suit claiming that Australia's state pension discriminates against them because their life expectancy is much lower than non-Aboriginals. Recalling the curves and discussions on longevity from Chapter 2, most Aboriginals live to their mid-70s, while non-Aboriginal Australians live into their 80s and 90s. Their "investment return" from their national pension scheme—a calculation I'll explain in much more detail in the next section—falls short of the comparable numbers for non-Aboriginals. The case, which garnered much international attention, was (at the time of writing) pending before the Federal Court of Australia, and similar concerns are being expressed elsewhere around the world.

Economically and practically speaking, there might be a point at which it makes perfect sense for a participant in a national longevity risk pooling scheme—that is, someone or some group with a noticeably low life expectancy in retirement—to completely opt out of the plan. There is a point at which they aren't benefiting at all from the longevity risk pooling within national pensions. So, let's delve into the cash flow ins and outs of those plans to get a better sense of the point at which opting out is the optimal strategy.

National Pension Pools: Fair IRRs

As noted in the introductory chapter of this book, most developed countries have national pension schemes that force all working citizens to contribute savings to a collective fund, what I have called the longevity risk pool. As you know by now, this sort of actuarially adjusted scheme

and Finance, where a more technical approach and discussion is provided. It is cited as Milevsky (2020).

was first implemented by (the Protestant) Reverend Alexander Webster and his Scottish colleagues in 1744, and then made its way around the world in the centuries that followed. But instead of the local church or university managing a plan, it is now mandated on a national scale. To be clear, national pensions are distinct from corporate retirement plans that are voluntary and whose generosity is at the discretion of the employers. Nevertheless, both types of plans—sponsored by governments or corporations—pool longevity risk in one manner or another.

For example, in Canada, the national pension scheme is a mandatory employment-based plan called the Canadian Pension Plan (CPP). In this plan, all workers must contribute (at most) $5800 per year in CPP premiums in exchange for (at most) $15,700 per year beginning at 65, the official retirement age. These amounts are less than the $100,000 that Colin (or anyone else) might need for retirement, but if he lived in Canada (in the twenty-first century), and was entitled to full CPP benefits, then all he would require from his personal assets is $84,300, to make up the difference. But that's not the point here. My focus here is on the communal or national pension fund and the fact that approximately $15,700 of his retirement income comes from a longevity risk pool. After all, if Colin lives for 30 years in retirement while his identically aged neighbour Alexander lives for 10 years, then implicitly Colin is being subsidized by Alexander. That's the nature of pooling.[2]

Now, regardless of how much (or how little) you might judge $15,700 to be, a Canadian who makes these contributions from age 25 until age 65, that is 40 years, and then lives (only) to the average Canadian age of 85, that is 20 years in retirement, will earn an internal rate of return of 1% per year after inflation. I use the term "real interest rate" because the pension annuity is real and adjusted each year for inflation. For those readers who might be interested in confirming or replicating this IRR, note that the future value (FV) of the $5800 in contributions for 40 years is approximately $283,000 real dollars when accumulated at 1% per year. Notice also that the present value (PV) of the $15,700 benefit for 20 years, is also $283,000 when discounted at 1% per year. When the PV and FV are computed at a particular point in time—namely, at retirement age 65—and are matched at a given valuation rate—the 1%

[2] I'm abstracting somewhat from the current (2024) rules for the Canadian Pension Plan, and in particular the government added a future enhancement that complicates the calculation, but these are the correct ballpark figures.

in our case—then we declare that as the internal rate of return. Perhaps a technical point, but worth comparing with other comparable instruments.

Indeed, this 1% *real* number, which is lower than the 5% *nominal* number used in Chapter 3, or the 4% *nominal* number used by the trustees of the Presbyterian fund, is slightly lower than risk-free interest rates, but is effectively guaranteed by the government. So, the 1% real interest isn't great, but it is decent. More broadly, the paternalistic calculus is to force all citizens to participate, since many would be unlikely to do so on their own. Again, this is for a stylized individual who works for 40 years and lives 20 years in retirement.

Alas, the challenge or dilemma—and core of this chapter—is what happens to those unhealthy retirees (think Aboriginal Australians) who don't spend 20 years in retirement and don't live to age 85, as in the above example. Those who are unfortunate to live just 10 years longer to age 75, for example, will obviously earn a much lower internal rate of return (IRR). After all, the payments are still being made for 40 years, but the income lasts for only 10 years. One doesn't require sophisticated MBA mathematics to arrive at this insight. Indeed, contributing a total of $232,000 over one's life, which is 30 × $5800 in exchange for only $157,000 = 10 × $15,700, isn't a good investment, especially considering the individual was forced to participate. In a later section of this chapter, I provide yet another perspective on how much the "longevity winners" gain at the expense of the "longevity losers" and perhaps how much more those with a lower life expectancy should receive. For now, I begin by hinting that there should be some compensation to those (perhaps Aboriginals) who are extremely unlikely to ever hit the population averages.

Now, again, in defence of any national pooling system, the conventional and centuries-old response by pension economists—and you have heard this a few times in this book already—is that for every unlucky person who only lives to age 75 there is another lucky one who lives to a ripe old age of 95. Moreover, defenders argue, that is the nature of pooling. Winners and losers are only known at the end, *ex post*, but everyone benefits from longevity insurance *ex ante*.

Yet a deeper dive into the numbers, using the methodology and language of the 1744 Presbyterian scheme, reveals very large discrepancies indeed. Again, my (policy) point for this chapter is that we might have to rethink fair longevity risk pooling. In the next few pages I'll attempt to explain the magnitude of the transfer from those with a known longevity deficit to those with a known longevity surplus.

Wealth Transfers: From Less to More Fortunate

Once again I'll return to Colin and Anne from Chapter 3, who are members of the same (hypothetical) national pension plan that pools longevity risk with everyone, together. Assume that Anne is about to retire at 65 and is entitled to a pension annuity of $25,000 per year, paid monthly for the remainder of her life. This is the maximum she can receive from her scheme after having worked the requisite number of years. Lucky for her, the pension annuity payments are adjusted or increased every year for the cost of living or price inflation as measured by some national index. However, the income will cease upon her death. It contains no cash value or liquidity provisions, nor can she bequeath the pension income to her children, grandchildren, or loved ones.

Now, her husband Colin, who coincidentally was born in the same year as Anne, is also about to retire at age 65 and entitled to the same pension annuity of $25,000 per year. He too has contributed the maximum to the pension scheme or retirement program, perhaps explicitly by having a fraction of his paycheque withheld or implicitly via the income tax system. The details of how Colin and Anne paid for their pension annuity are unimportant at this juncture. The key element, though, is that over his working life Colin has contributed the same amount of money as Anne to the pension system.

I now get to the remonstrance of the matter. There is one important difference between Colin and Anne, in addition to their gender. Regrettably, Colin has a medical life expectancy of 10 years and is rather sickly, whereas Anne is in perfect health with a corresponding life expectancy of 30 years. You know from the argument just made in the prior section, when I mentioned the Canadian Pension Plan that Colin loses relative to Anne. How much is the loss?

Remember that despite Colin's poor health—and the financial fact he contributed the exact same amount to the retirement program—he isn't entitled to any more income from his pension annuity compared to Anne, because (again) retirement programs aren't underwritten or adjusted for the individual's health status. In fact, most government national pension programs around the world like the CPP are unisex as well as gender neutral. If you contributed (the maximum amount) into the system, you will receive the same benefit regardless of your sex or health status or the amount of time you are expected to live in retirement.

Now you might *believe* this is completely normal, fair, and socially progressive and that any pension scheme should not discriminate in any way—and I have no intention of arguing with you. Rather, I would like to focus attention on the financial implications of a heterogenous or "blind" risk pool to gender, health, demographics, and genetic predisposition. It's quite clear that Colin's short life expectancy of 10 years implies that he will be receiving (much) less income compared to Anne. Moreover, if the retirement program is designed to neither make nor lose money in the long run, that is, to be actuarially balanced, then the Colins of the world are subsidizing the Annes of the world.

Once again, think back to the Aboriginal Australians, this statement isn't meant to be controversial or debatable. It's the exact opposite of life insurance. In this case, people who die early into their retirement years, subsidize those who live a long and healthy life. If the wealth transfer does come up in polite conversation, the justification provided for the subsidy is that, notwithstanding the current unfairness, it might be desirable to help undo or repair historical inequities. Alas, that is the nature of all social programs and other public policies designed to provide assistance. Others will argue in defence of this demographic blindness that in practice most real-world (not hypothetical) Annes are entitled to only $15,000 (not $25,000) per year while most real-world Colins are receiving $30,000 (not $25,000) in pension income. Indeed, perhaps Anne was forced to leave work and the labour force over multiple years to give birth, raise a family, or take care of an infirm relative. In other words, the transfer isn't that large.

Again, I'm not arguing that the transfer is unfair or unprecedented. Most income tax systems are progressive, which means that high-income taxpayers subsidize low-income taxpayers. But in this case, the transfer is from those with shorter lives to those with longer lives. Once again there is no absence of loud, emotional, and historical explanations offered to justify why Colin's pension annuity should be the exact same as Anne's and that her obviously longer life expectancy shouldn't affect her entitlements.

My intent here isn't to argue against the subsidy. Rather, what I would like to continue to do in the next few pages is develop a framework to help quantify the magnitude of the subsidy from Colin to Anne and illustrate that it's quite large under reasonable parameters...and perhaps

Colin deserves a slightly larger pension (more than $25,000) even if he contributed the exact same as Anne.

Life Annuities: Quantify the Wealth Transfer

To get to the essence of the pooling and subsidies from a financial perspective, the next questions I ask are as follows: How much would Colin have to pay in the open (i.e., retail) market to acquire a pension annuity of $25,000 per year? How much would Anne have to pay? You might think that only pension plans can offer these life annuity promises or guarantees, but they are in fact available for purchase from most (corporate) life insurance companies, no different than life insurance, home insurance, car insurance, or any other form of insurance they sell to consumers. Recall the discussion in Chapter 11, of insurance companies and associations that objected to national pension schemes because they effectively competed with their products.

Now, in practice, the market price or charge for the retail (versus bulk) life annuity would depend on many factors, including the uncertainty around both Colin's and Anne's life expectancy. So, to keep things very simple (at first), I'll assume that Colin will live for exactly 10 years longer and Anne will live for exactly 30 years longer. In other words, I'll assume it's not an average but a reality. And, for the actuarial and insurance readers who are unhappy with this deterministic assumption, perhaps I can assuage your concerns by assuming the pension annuity is a "term-certain annuity" and will only pay Colin for 10 years and Anne for 30 years.

Either way, here are the financial facts of life. Colin would be charged $212,750 at 65 and Anne would be charged $487,250 for the same exact pension annuity. These numbers are based on a 3% effective (real) annual interest rate, no default risk, and ignoring taxes, but does not require much else in terms of assumptions or parameters.

Stated differently, the present value of $25,000 per year for 10 years is exactly $212,750. But for 30 years of cash flow the present value is $487,250. This is the exact same calculation that I used earlier. And, thinking in terms of a collective pool, the market cost of their combined pension annuity entitlement is $212,750 + $487,250 = $700,000. So, if—and this is a big if—the scheme is properly funded (i.e., actuarially balanced), it should have exactly $700,000 set aside in reserves to pay pension annuities for Colin and Anne at the exact point or moment of retirement.

OK. Spoiler alert here: Very few pension schemes have anywhere near the $700,000 set aside that is needed to pay all the guaranteed pension annuities, which is yet another reason for reform. But I'm getting way ahead of myself here and it's not relevant to the economic value of Anne's and Colin's pension entitlements. Proper funding of government and state pensions is more of a political issue really, and far outside the mandate of this book.

To be clear, real-world insurance companies will not charge $212,750 and $487,250 to Colin or Anne for true life annuities. First and foremost, these companies have to make profits, so they would mark up or load the price of insurance no differently than how it is done for the retail versus the wholesale cost of coffee. More importantly, insurance companies have to budget provisions to cover the uncertainty, including the risk of how long their annuitants will live.

The first takeaway from these two numbers is as follows: Colin is transferring $137,250 to Anne. This is a subsidy amounting to 64.5% of the hypothetical value of Colin's pot of money. Why? Again, the pension system should have $700,000 set aside for both of them, of which $487,250 is needed for Anne and only $212,750 is needed for Colin. And yet they both contributed the same amount of money to the system—presumably a total of $350,000 × 2 = $700,000 over the course of their working lives. Colin contributed $350,000 and is getting something worth $137,250 less. Anne contributed $350,000 and is getting something worth $137,250 more; namely, the market price of the pension annuity she will receive for the rest of her life. Table 12.1 provides a summary.

Table 12.1 Longevity risk pooling: hypothetical winners and losers

	Colin	*Anne*
Annuity Income	$25,000 p.a.	$25,000 p.a.
Life Horizon	10 years	30 years
Annuity Value (A)	$212,750	$487,250
Contributions (B)	$350,000	$350,000
Transfer A – B	–$137,250	+$137,250
Subsidy (A – B) / A	–64.5%	+28.2%

Source Author's calculations

Yes, of course, these are straw men and women. The numbers assume a pension system with no spousal or survivor benefits and a (somewhat extreme) 20-year gap in life expectancy between the two. More importantly, my simple tale assumes that Colin and Anne die precisely at the age of their life expectancy, which presumes the absence of any longevity risk underlying Chapter 2. In reality, Colin might live beyond age 75 (or he might die even earlier), and Anne might not make it to age 95 (or she might live even longer). Under those outcomes, the subsidy might be smaller (or even larger).

I really won't know precisely *who paid whom* until they both pass away, or as economists would say, *ex post*. But the *ex ante* reality is that there is a large gap between the expected present value of the benefits they receive even though they have paid similar amounts into the retirement program. In fact, if there are more Annes than Colins in the program, the subsidy provided by Colin is even greater. Vice versa, if the retirement program is stacked with many more Colins relative to fewer Annes, the subsidy provided by any one Colin isn't as large and might yet be another reason for why this issue isn't a big deal.

My point is that whenever you mix (i.e., pool) heterogeneous people with different longevity prospects into one big pension scheme in which everyone gets a pension annuity for the rest of their natural lives, there will be clear winners and losers *ex ante*. And the sums being transferred from the latter to the former can be substantial.

But what happens if I incorporate longevity risk or horizon uncertainty into our straw neighbours? Well, as I noted, there is a small probability that Colin lives for more than 10 years beyond age 75 and/or that Anne dies before age 95, and the exact terminal horizon is unknown in advance. In that case, the *ex post* transfer of wealth from Colin to Anne would be less than 60% of the value of his pension annuity. In fact, there is a (very) small probability that Colin actually outlives Anne and the *ex post* subsidy is in reverse and she paid for him. Again, I won't know until all the Annes and Colins are dead.

But here is my next point. The pension annuity they are entitled to provides them with more than just a periodic cash flow or income; it provides them with longevity insurance. Moreover, the value or benefit of any type of insurance can't be quantified in terms of what might happen on average. It must account for the tails or edges of the distribution.

To the issue of fairness, consider the fact that we all happily buy car insurance, home insurance, and even life insurance in which the expected discounted value of the monetary benefits is less than what we pay. And yet, it makes perfect financial sense to do so. Why? It's about protection. Allow me to describe this in a way that a financial economist would prefer. Would you be willing to receive less than what you paid *on average*, but where the big payouts occurred in the states of nature in which you need and subjectively value them the most? Most economists would answer with a resounding yes. *On average* you don't die from the flu, you are never in a car accident, and your house never floods. *On average* bad things never happen and insurance is wasted. But you don't buy these things for the average. You buy them for the tails.

Yes, this might all sound rather technical and we re-enter the realm of insurance economics and the utility value of risk mitigation, which was explained in Chapter 3. Nevertheless, going back to Anne and Colin. As mentioned earlier, their income annuity entitles them to more than a term-certain annuity for 30 and 10 years, respectively. They have acquired longevity insurance that protects them in the event they live longer than average. Colin would rather be "pooled" with people like him who share the same longevity risk profile. But perhaps even Colin is willing to be pooled with people like Anne if the alternative is no pooling at all.

Can the Volatility of Longevity Justify the Subsidy?

So, who values this insurance more? Or is the insurance benefit symmetric? Stated differently, could Colin be gaining more (in utility) from pooling with Anne, even if he is losing on an expected present-value basis? In theory, yes, Colin could be winning (economic utility) even if he appears to be losing (dollars and cents).

Why? In a nutshell, Colin's volatility of longevity is greater. He has a short life expectancy, but relatively speaking, the range of how long he might actually live, expressed as a percentage, is actually larger than Anne's. Intuitively, the notion of Colin living 30 years instead of 10 years is equivalent to a 200% "shock" to expectations. It's unlikely, but in the realm of possibility. In contrast, Anne who is expected to live 30 years will never experience a 200% shock. This would imply she lives 90 more years (from age 65) to the age of 155. It simply won't happen. The odds are zero. Ergo, Colin's volatility of longevity (in relative percentage terms) is higher than Anne's. That's biostatistics. But from an insurance economics point of view, it implies that Colin values the risk pooling benefits of the pension more than Anne does. If you need a refresher, go back to Fig. 2.1. Think of Group A as the Annes of the world, and Group B as representing the Colins.

What is the bottom line for all of this? Well, for the purposes of longevity risk pooling the risk-adjusted transfer from Colin to Anne isn't as large as the expected dollars indicate because Colin's volatility of longevity is larger than Anne's. Does it completely negate the fact that Colin will be receiving payments for 10 years (on average) and Anne will be earning them for 30 years (on average)? No. Not completely. But hopefully at this point of the narrative you have a much better and clearer picture of not only how longevity risk pooling works but also who the winners and losers might be and what that might depend on.

Should the Short Lived Get Bigger Pensions?

In this section I offer one final insight or perspective into the problem, by examining the payout multiple—a concept familiar from Chapters 5, 6, and 7—that might solve the longevity inequality problem. Table 12.2 provides an indication of how much (more) the pension entitlement for those who are short-lived might be, under two different system-wide interest rate assumptions (1% real versus 4% real) and how those retirement pension payouts would compare across longevity profiles. Before you delve too deeply into these numbers, I should caution that they are hypothetical, replete with assumptions, and certainly not meant to account for any national pension scheme. Especially, on an after-tax basis, there are claw-back benefits for high-income earners. Nevertheless, here is an explanation of some key numbers in the table.

We begin with a national pension system that is set up based on a 1% real (inflation adjusted) valuation rate and is built to actuarially balance contributions against benefits. In such a system, if everyone contributed \$1 per year for 40 years, and then lived exactly 20 years in retirement, their pension income should be \$2.71 per year for life. If they have contributed \$10,000 per year instead of \$1 per year, their pension would be scaled accordingly to \$22,710. Likewise, if their contribution was \$5800 (per year) the benefits should be 2.71 times \$5800 which is approximately \$15,700 (per year.) These last two numbers should sound familiar, as they echoed the CPP mentioned earlier in this chapter. Likewise, the multiple of 2.71 might remind readers of the multiple (of 3.81) in the Presbyterian annuity fund, although there is a bit of apples and

Table 12.2 Work and contribute $1 per year to pension for 40 years

Retirement longevity (in Years)	*Fair pension payout multiple*		*Premium or discount vs. average longevity*	
	IRR of 1%	*IRR of 4%*	*IRR of 1%*	*IRR of 4%*
10	5.16	11.72	*+90.5%*	*+67.6%*
15	3.53	8.55	*+30.2%*	*+22.2%*
Average = 20	2.71	6.99	*N.A*	*N.A*
25	2.22	6.08	*−18.1%*	*−13.0%*
30	1.89	5.50	*−30.1%*	*−21.4%*
35	1.66	5.09	*−38.6%*	*−27.2%*

Source Author's calculations

oranges in the comparison since they apply to different people at different stages of their life. Nevertheless, how to interpret and use the multiple should be clear by now.

Let's examine the other (top and bottom) numbers in the same Table 12.2 Assume that Colin (once again), who is a representative retiree with low longevity, is only anticipated to live 10 years instead of 20 years. Yes, this isn't quite the same Colin from the prior sections or the Colin in Chapter 3, but notice what the fair pension multiple should be in the upper left-hand corner of Table 12.2: $5.16.

What this means is that for the system to be fair to Colin, under a 1% valuation interest rate, then he should be getting 5.16 times his annual contributions as his retirement pension, and not the 2.71 given to those with a 20-year retirement longevity. Moreover, this is 90.5% more than the payout multiple for the average person in the plan. The intuition here is quite simple. He contributed the same amount ($1) and the same time (40 years) as everyone else in the system, but is only living for 10 (versus 20) years. Ergo, he must be compensated with a high pension multiple. How much higher? Well, if the system valuation rate is 1%, then he should get 90% more than the "average" participant. This, in a nutshell, might be exactly what the Aboriginal Australians are arguing for, as mentioned at the start of this chapter.

Now, to the other side of this coin. If the Colins of the world with a retirement longevity of 10 years should be getting 90% more via higher payout multiples, then there is a group that should be getting lower multiples, or less than average. At the other extreme is Anne with a longevity of

35 years. Again, this is entirely hypothetical to make a point in clear and easy to understand terms. Anne must have her payout multiple reduced from the 2.71 average to 1.66 per dollar, under a 1% system valuation rate. What does this mean in practice? Lower pension income compared to Colin. In fact, it should be 38.6% lower than those with average longevity.

All of the above is based on a system valuation rate of 1%, which is the real after inflation rate of return that the national pension system is targeting, and a rather low number considering the recent rise in world interest rates. If, in contrast, the valuation rate is raised to 4% real, which is quite high, then the numbers change but the intuition remains exactly the same. The average payout multiple should increase from $2.71 to $6.99, because money earns more when rates are 4% versus 1%. This is why Colin (with 10 years to go) should get a much higher multiple of $11.72, which is 67.6% more than the average payout for those with 20 years longevity.[3]

So, which is it? Should Colin's pension be 90.5% higher than the system-wide average pension, or only 67.6% more? Well, think of those two numbers as a reasonable range for people who longevity is likely to be 10 years less than the "average." And, while it's quite dangerous to banter about the word average so many times in one paragraph, hopefully the main message is crystal clear in all cases. If you are part of an identifiable group with lower longevity prospects, your pension annuity should be higher. Period.

OK then, is it realistic to identify a definite group of people with a retirement longevity of (only) 10 years versus another group with a retirement longevity of (at least) 35 years? Is it that clear and easy to separate these two groups? The next few pages examine the evidence.

[3] While the internal rate of return (IRR) was explained earlier in this chapter, for those readers who want to replicate the numbers in Table 12.2, here is an intermediate step. Under a 4% valuation rate, the future value of the $1 for 40 years is $95.03 at retirement. Think of that is the accumulated value of the pension fund. So, the present value (PV) of the payout in dollars, must also equal $95.03. Indeed, the PV of $11.72 for 10 years (Colin) is equal to the PV of 5.09 for 35 years (Anne), namely $95.03. Indeed, all the present values must be the same, similar to the arguments made around Table 12.1. In some sense the message is the same, but the examples are different.

The Longevity Gap

Let's return to the growing body of empirical research by demographers, statisticians, medical researchers, epidemiologists, and insurance actuaries pointing to a growing gap in longevity prospects within subsegments of the population, significantly when they diverge in matters of wealth and income. The most widely cited article linking income to life expectancy (in the US) is the study by Chetty et al. (2016), and a similar article in Canada is by Milligan and Schirle (2021), using CPP data. Other researchers have looked at non-financial factors for the stochastic longevity gap vis-à-vis the implications for pension plans, both from a theoretical and empirical perspective.

That literature—on the heterogeneity of longevity and the impact on pension fairness—is vast and growing.[4] And, while a long footnote is clearly not a proper review of the literature, the underlying messages are identical. Namely, the most crucial empirical takeaway is the existence of an identifiable group within society that will not live as long as the fortunate ones. Whether it is for purely genetic reasons—of which some claim a third of longevity can be attributed to—or to factors that are somewhat in a person's control, the fact is we can partially identify the longevity winners and losers, in advance.[5]

Yet most national and corporate pension schemes are all pooled together in one sizeable longevity risk-sharing fund or pool. At what point—and for what gap in the force of a mortality curve—does it no longer make sense to be a member of such a longevity risk-sharing pool? These matters border on philosophical questions related to ageing and ethics, as noted in Wareham (2022), as well as by the previously mentioned philosophers, Kekes (1997) and Mothersill (1999).

[4] This includes Aburto et al. (2022), Alexander (2022), Ayuso et al. (2017), Baxter (2019), Bilal et al. (2019), Bravo et al. (2023), Dwyer-Lindgren et al. (2022), Coppola et al. (2022), Couillard et al. (2021), Currie (2018), Denworth (2022), Dudel and van Raalte (2023), Finegood et al. (2021), Himmelstein et al. (2022), Kinge et al. (2019), Li and Hyndman (2021), Lin et al. (2017), Milligan and Schirle (2021), with particular emphasis to Canada, Mackenbach et al. (2019), Mosher (2023), Pitacco (2019), Perez-Salamero Gonzalez et al. (2022), Sanzenbacher et al. (2019), Shi and Kolk (2022), Shi et al. (2023), Simonovits and Lackó (2023), Sloan et al. (2010), Strozza et al. (2022), Walczak et al. (2021), Wen et al. (2020), Wilmoth et al. (2011), and Woolf (2023).

[5] See, for example, Fleurbaey et al. (2014) and Ponthiere (2023) for some economic articles on how to balance these tensions and in-essence transfer wealth from those who "die late" to this who "die earlier."

Ethics of Pension Equity

The scholarly and practical concerns regarding pension equity and the transfer of benefits from short-lived to long-lived retirees go beyond technically trained financial economists and insurance actuaries. In fact, even academic philosophers have taken a nascent interest in the ethics of (what mathematicians label) stochastic longevity. A sample of this literature can be found in the *Cambridge Handbook of the Ethics of Ageing*, published in 2022, which includes contributions from 20 *bona fide* moral philosophers and bioethicists. The introduction to this collection of articles penned and edited by C.S. Wareham notes that there is a gap in the literature regarding the focused treatment of ageing (back to US style) in the field of philosophical ethics. Questions around social justice and the need to balance the interests of different generations and age cohorts are taking centre stage among philosophers and even financial economists.[6]

In a rather provocative article published towards the end of the twentieth century, the philosopher John Kekes noted the longevity gender gap, which he described as one-tenth shorter for males than females, and labelled it as an ***unjustified inequality*** that should be remedied. Appealing to the egalitarian views of John Rawls and Thomas Nagel, Kekes writes that since men have shorter lives than women (on average, of course), "they are less likely to benefit after retirement from social security payments." Of course, that fact doesn't require a financial economist and isn't controversial. But Kekes then goes on to state—as a philosopher—that it is unjust from an egalitarian point of view that men should be required to subsidize the health and wealth of women in their declining years and that policy should suggest pension contribution rates should be lowered on men and increased on women. These egalitarian arguments, which were written over a quarter of a century ago, would apply equally to the non-gendered groups. Kekes himself might have been creating a strawman argument since he also writes, "Redistribution ought to aim to equalize the life expectancy of men and women by making men have longer and women shorter lives," which is obviously in jest. But the point here is simply that issues of pension equity that emanate from mortality

[6] See in particular the collection edited by C.S. Wareham (2022). See also the article by Timothy Johnson (2015) who argues that reciprocity is the foundation of financial economics, and emphasized cooperation as part of the market-clearing process.

heterogeneity have captured the interest of (even) the most humanistic area of the social sciences, namely, the philosophers.[7]

Echoing similar themes during the same late twentieth-century time period, the president of the American Philosophical Association and Columbia University Professor of Philosophy Mary Mothersill (a Canadian, who died in 2008) made the following introductory comment: "Bad consequences flow from the gerontologist's neglect of philosophy." No doubt Professor Mothersill was using the term gerontology as a catch-all phrase for the non-philosophical studying of ageing. She mentions a number of specifically philosophical issues that arise when reflecting on old age, including the question: "Does value, positive or negative, attach to longevity?" Mothersill concludes her presidential remarks by citing David Hume (died at age 65), who in the last few months of his life (1776) wrote the following: "I consider, besides, that a man of 65 by dying cuts-off only a few years of infirmities."[8]

In sum, the question of pension equity—which is at the core of this chapter—deserves the attention of the broader community, beyond narrow technocrats. Echoing Kekes, moral philosophers have recently gone so far as to question whether women's lifespans are a grave injustice and an unfair advantage that must be corrected within the pension system. Others note that even fully funded pension systems (in contrast to PAYGO schemes) involve some level of redistribution from young to old, which might go against the principles of distributive justice. So, while the above-cited philosophers focused on the longevity gender inequality, the same arguments can be made for the Indigenous, the poor, or other groups in which longevity risk is not shared equitably. Axel Gosseries (2022) makes it very clear when he says that his focus is to examine what, if anything, "long-lived people owe to short-lived people with what a given generation owes to the previous and/or the next ones."[9]

[7] Kekes, J. (1997).

[8] Mothersill, M. (1999).

[9] See Brennan, S. (2022), as well as Gosseries, A. (2022).

Final Fund Epilogue

While the focus of this book has been on the origins and roots of the Scottish minister's fund for widows and orphans, and in particular the archival record from the first decade of its operations, this is likely a good time to mention an interesting development that took place in 1862, over a century later.[10] As noted in earlier chapters and emphasized in this final one, a contribution or premium basis that is independent of age leads to inequities among members. In a pure *life insurance* scheme, with the exact same level of death benefits, younger members should pay less than older ones. And, vice versa, in a pure *life annuity* scheme with equal pension benefits, older members should pay less than younger ones. The single premium annuity factor is a declining function of age. When it came to the Scottish scheme, which had elements of both life insurance and life annuity benefits, the older ministers and professors should have paid more in annual premiums (see Chapter 7). Alas, in the year 1862, the trustees of the scheme asked a proper actuary to review the details of the plan. By then, all of the original trustees were long gone and the professional actuary was a recognized and respected expert on these matters. The actuary of the plan determined that the scheme rules as formulated—equal premiums regardless of age—were no longer viable. Instead, and from that time onward, every single member would be treated individually. And so, a small but meaningful part of Alexander Webster's heterogenous pool of altruistically blended young and old church brethren, died quietly on Whitsunday 1862.

Sources and Additional References

Aburto, J. M., Basellini, U., Baudisch, A., & Villavicencio, F. (2022). Drewnowski's Index to Measure Lifespan Variation: Revisiting the Gini Coefficient of the Life Table. *Theoretical Population Biology, 148*, 1–10.

Alexander, M. (2022). *Decomposing Dimensions of Mortality Inequality*.

Ayuso, M., Bravo, J. M., & Holzmann, R. (2017). Addressing Longevity' Heterogeneity in Pension Scheme Design. *Journal of Finance and Economics, 6*(1), 1–21.

[10] This information is an amalgamation from Sibbett (2014) as well as online material available from the Institute and Faculty of Actuaries (IFoA) in the UK, under historical collections.

Baxter, S. (2019). *Understanding Socio-Economic Differences in Pension Plan & Insurer Longevity.*
Bilal, U., Cainzos-Achirica, M., Cleries, M., Santaeugenia, S., Corbella, X., Comin-Colet, J., & Vela, E. (2019). Socioeconomic Status, Life Expectancy and Mortality in a Universal Healthcare Setting: An Individual-Level Analysis of >6 Million Catalan Residents. *Preventive Medicine, 123*, 91–94.
Bravo, J., Ayuso, M., Holzmann, R., & Palmer, E. (2023). Intergenerational Actuarial Fairness when Longevity Increases: Amending The Retirement Age. *Insurance: Mathematics and Economics, 113*, 161–184.
Brennan, S. (2022). To Grandmother's House We Go: On Women, Ethics, and Ageing. In *The Cambridge Handbook of the Ethics of Ageing.*
Chetty, R., Stepner, M., Abraham, S., Lin, S., Scuderi, B., Turner, N., Bergeron, A., & Cutler, D. (2016). The Association Between Income and Life Expectancy in the United States, 2001–2014. *JAMA, 315*(16), 1750–1766.
Coppola, M., Russolillo, M., & Simone, R. (2022). On the Evolution of the Gender Gap in Life Expectancy at Normal Retirement Age for OECD Countries. *Genus, 78*(1), 27.
Couillard, B. K., Foote, C. L., Gandhi, K., Meara, E., & Skinner, J. (2021). Rising Geographic Disparities in US Mortality. *Journal of Economic Perspective, 35*(4), 123–146.
Currie, J. M. (2018). Inequality in Mortality Over the Life Course: Why Things Are Not as Bad as You Think. *Contemporary Economic Policy, 36*(1), 7–23.
Denworth, L. (2022). People in Republican Counties Have Higher Death Rates Than Those in Democratic Counties. *Scientific American.*
Dudel, C., & van Raalte, A. A. (2023). Educational Inequalities in Life Expectancy: Measures, Mapping, Meaning. *Journal of Epidemiology and Community Health, 77*(7), 417–418.
Dwyer-Lindgren, L., Kendrick, P., Kelly, Y. O., Sylte, D. O., Schmidt, C., Blacker, B. F., & Mokdad, A. H. (2022). Life Expectancy by County, Race, and Ethnicity in the USA, 2000–19: A Systematic Analysis of Health Disparities. *The Lancet, 400*(10345), 25–38.
Finegood, E. D., Briley, D. A., Turiano, N. A., Freedman, A., South, S. C., Krueger, R. F., Chen, E., Mroczek, D. K., & Miller, G. E. (2021). Association of Wealth with Longevity in US Adults at Midlife. *JAMA Health Forum, 2*(7).
Fleurbaey, M., Leroux, M.-L., & Ponthiere, G. (2014). Compensating the Dead. *Journal of Mathematical Economics, 51*, 28–41.
Gosseries, A. (2022). Ageing, Unequal Longevities and Intergenerational Justice. In *The Cambridge Handbook of the Ethics of Ageing.*
Himmelstein, K. E. W., Lawrence, J. A., Jahn, J. L., Ceasar, J. N., Morse, M., Bassett, M. T., Wispelwey, B. P., Darity, W. A., Jr., & Venkataramani, A. S. (2022). Association Between Racial Wealth Inequities and Racial Disparities

in Longevity Among US Adults and Role of Reparations Payments, 1992 to 2018. *JAMA Network Open, 5*(11), e2240519.
Johnson, T. C. (2015). Reciprocity as a Foundation of Financial Economics. *Journal of Business Ethics, 131*(1), 43–67.
Kekes, J. (1997). A Question for Egalitarians. *Ethics, 107*(4), 658–669.
Kinge, J. M., Modalsli, J. H., Overland, S., Gjessing, H. K., Tollanes, M. C., Knudsen, A. K., Skirbekk, V., Strand, B. H., Haberg, S. E., & Vollset, S. E. (2019). Association of Household Income With Life Expectancy and Cause-Specific Mortality in Norway, 2005–2015. *JAMA, 321*(19), 1916–1925.
Li, H., & Hyndman, R. J. (2021). Assessing Mortality Inequality in the U.S.: What Can Be Said about the Future? *Insurance: Mathematics and Economics, 99*, 152–162.
Lin, T. Y., Chen, C. Y., Tsao, C. Y., & Hsu, K. H. (2017). The Association Between Personal Income and Aging: A Population-Based 13-Year Longitudinal Study. *Archives of Gerontology and Geriatrics, 70*, 76–83.
Mackenbach, J. P., Valverde, J. R., Bopp, M., Bronnum-Hansen, H., Deboosere, P., Kalediene, R., Kovacs, K., Leinsalu, M., Martikainen, P., Menvielle, G., Regidor, E., & Nusselder, W. J. (2019). Determinants of Inequalities in Life Expectancy: An International Comparative Study of Eight Risk Factors. *Lancet Public Health, 4*(10), e529–e537.
Milevsky, M. A. (2020). Swimming with Wealthy Sharks: Longevity, Volatility and the Value of Risk Pooling. *Journal of Pension Economics and Finance, 19*(2), 217–246.
Milligan, K., & Schirle, T. (2021). The Evolution of Longevity: Evidence from Canada. *Canadian Journal of Economics/Revue canadienne d'économique, 54*(1), 164–192.
Mosher, J. (2023). *Mortality and the Provision of Retirement Income*. OECD.
Mothersill, M. (1999). Old Age. *Proceedings and Addresses of the American Philosophical Association, 73*(2), 7–23.
Perez-Salamero Gonzalez, J. M., Regulez-Castillo, M., Ventura-Marco, M., & Vidal-Melia, C. (2022). Mortality and Life Expectancy Trends in Spain by Pension Income Level for Male Pensioners in the General Regime Retiring at the Statutory Age, 2005–2018. *International Journal for Equity in Health, 21*(1), 96.
Pitacco, E. (2019). Heterogeneity in Mortality: A Survey with an Actuarial Focus. *European Actuarial Journal, 9*(1), 3–30.
Ponthiere, G. (2023). An Insurance Against a Short Life. In *Allocating Pensions to Younger People: Towards a Social Insurance Against a Short Life* (pp. 39–72). Springer.
Sibbett, T. (2014). History of Insurance. *Wiley StatsRef: Statistics Reference Online*. https://doi.org/10.1002/9781118445112.stat04448

Sanzenbacher, G. T., Webb, A., Cosgrove, C. M., & Orlova, N. (2019). Rising Inequality in Life Expectancy by Socioeconomic Status. *North American Actuarial Journal, 25*(Suppl. 1), S566–S581.

Shi, J., Aburto, J. M., Martikainen, P., Tarkiainen, L., & van Raalte, A. (2023). A Distributional Approach to Measuring Lifespan Stratification. *Population Studies (Camb), 77*(1), 15–33.

Shi, J., & Kolk, M. (2022). How Does Mortality Contribute to Lifetime Pension Inequality? Evidence from Five Decades of Swedish Taxation Data. *Demography, 59*(5), 1843–1871.

Simonovits, A., & Lackó, M. (2023). A Simple Estimation of the Longevity Gap and Redistribution in the Pension System. *Acta Oeconomica, 73*(2), 275–284.

Sloan, F. A., Ayyagari, P., Salm, M., & Grossman, D. (2010). The Longevity Gap Between Black and White Men in the United States at the Beginning and End of the 20th Century. *American Journal of Public Health, 100*(2), 357–363.

Strozza, C., Vigezzi, S., Callaway, J., Kashnitsky, I., Aleksandrovs, A., & Vaupel, J. W. (2022). *Socioeconomic Inequalities in Survival to Retirement Age: A Register-Based Analysis.*

Walczak, D., Wantoch-Rekowski, J., & Marczak, R. (2021). Impact of Income on Life Expectancy: A Challenge for the Pension Policy. *Risks, 9*(4), 65.

Wareham, C. S. (2022). *The Cambridge Handbook of the Ethics of Ageing.* Cambridge University Press.

Wen, J., Kleinow, T., & Cairns, A. J. G. (2020). Trends in Canadian Mortality by Pension Level: Evidence from the CPP and QPP. *North American Actuarial Journal, 24*(4), 533–561.

Wilmoth, J. R., Boe, C., Barbieri, M., Crimmins, E., Preston, S., & Cohen, B. (2011). Geographic Differences in Life Expectancy at Age 50 in the U.S. Compared with Other High-Income Countries. In *International Differences in Mortality at Older Ages* (pp. 333–366).

Woolf, S. H. (2023). Falling Behind: The Growing Gap in Life Expectancy Between the United States and Other Countries, 1933–2021. *American Journal of Public Health, 113*(9), 970–980.

Index

A
Actuarial Science, vii, 18, 21, 56, 72, 82, 85, 87, 102, 107, 145, 198
Archival Records, viii, 87, 174, 259
Asset allocation, 6, 20, 88

B
Biblical annuity, 206, 218

C
Charge and Discharge, 154
Choice architecture, 19
Church of Scotland, vii, ix, x, 14–16, 19–21, 72–75, 78, 79, 81, 85, 89, 96–98, 100, 101, 110, 112, 131, 132, 135, 139, 152, 157, 164, 166, 171–173, 176–179, 181–186, 198
Contribution rates, 79, 97, 109, 111, 129, 146, 177, 257

D
Decumulation, 50–53, 55, 57, 61, 63, 66
Defoe, Daniel (Author), 180, 181

E
Economic outcomes, 9, 10

F
Fair IRR's, 244
Financial management, 80
Financial security, 219
Fund management, 196
Fund projections, 160

G
Gompertz-Makeham mortality, 149

H
Halley, Edmond (Astronomer), 15, 16, 73, 77, 139, 145, 149, 152, 197, 215

M. A. Milevsky, *The Religious Roots of Longevity Risk Sharing*,
https://doi.org/10.1007/978-3-031-62403-2

Hutcheson, Francis (Philosopher), 19, 128–131

I
Interest rates, 59, 61, 62, 64, 134, 158, 162, 164, 193, 207, 217, 218, 246, 255

J
Judean genealogy, 211

K
King Jehoiachin, 209, 211, 213–215, 218

L
Life annuity, 6, 13, 14, 21, 51, 61, 62, 75, 88, 89, 128, 147–149, 152, 205, 214, 215, 221, 228, 232, 236, 249, 259
Longevity risk, vii, viii, 13, 17, 18, 21, 25–30, 42–47, 49, 50, 52, 54–56, 58–66, 71, 72, 74–76, 82, 89, 98, 99, 107, 118, 127, 137, 147, 166, 172, 178, 179, 186, 196, 206, 220, 229, 236, 238, 241, 243–247, 251–253, 258

M
Maclaurin, Colin (Mathematician), 15, 16, 19, 72, 73, 75, 77, 80, 92, 93, 108, 111, 126–128, 137, 138, 149, 161, 167, 173, 174, 178, 181, 188, 199, 214, 215
Mortality tables, 73, 198, 215, 216

N
National schemes, 89

P
Payout multiples, 146, 150, 151, 254
Pension equity & fairness, 18, 49
Pension fund management, 79
Pension system rankings, 5, 6
Presbyterian, ix, 13–21, 40, 49, 55, 61, 74–76, 79–81, 100, 101, 132, 135, 140, 145, 155, 166, 172, 181, 183, 198, 199, 205, 206, 233, 246, 253
Price, Richard (Actuary), 16, 73, 75, 77, 81, 88, 90, 149, 198, 199
Protestant Christian, 7–13, 15, 17, 18, 20, 37, 40, 42, 71, 229, 238

R
Reid, Thomas (Philosopher), 19, 72, 111, 130–132
Religious beliefs, 7, 9, 17, 31, 32, 78, 191, 198, 200, 201, 244
Retirement planning, 220
Risk pooling, 50, 60, 63, 252
Robertson, William (Principal), 19, 80, 97, 109, 112, 131, 132, 186

S
Scheme Trustees, viii, 19, 94, 99, 126, 129, 130, 163
Scottish enlightenment, vii, x, 14, 19, 71–74, 78, 80, 82, 96, 102, 112, 126, 128, 130–132, 186
Smith, Adam (Economist), vii, 19, 77, 128–130, 132, 147

V
von Bismarck, Otto, 12, 90

W

Wallace, Robert (Minister), 15, 16, 75, 77, 92, 108, 111, 127, 132–135, 137, 149, 161, 172–174, 178, 181, 185–188, 193, 199, 215

Wealth transfers, 247

Weber, Max (Sociologist), 11, 12, 79, 101

Widows & Children, 138

The manufacturer's authorised representative in the EU is Springer Nature Customer Service Centre GmbH, Europaplatz 3, 69115 Heidelberg, Germany. If you have any concerns regarding our products, please contact ProductSafety@springernature.com

Printed and bound by CPI Group (UK) Ltd, Croydon, CR0 4YY
01/10/2026
02225903-0009